A Short Introduction to Intuitionistic Logic

THE UNIVERSITY SERIES IN MATHEMATICS

Recent volumes in the series:

COMPLEX ANALYSIS AND GEOMETRY
Edited by Vincenzo Ancona and Alessandro Silva

ELLIPTIC COHOMOLOGY
Charles B. Thomas

ELLIPTIC DIFFERENTIAL EQUATIONS AND OBSTACLE PROBLEMS
Giovannia Maria Troianiello

FINITE SIMPLE GROUPS: An Introduction to Their Classification
Daniel Gorenstein

AN INTRODUCTION TO ALGEBRAIC NUMBER THEORY
Takashi Ono

MATRIX THEORY: A Second Course
James M. Ortega

PROBABILITY MEASURES ON SEMIGROUPS: Convolution Products, Random Walks, and Random Matrices
Göran Högnäs and Arunava Mukherjea

RECURRENCE IN TOPOLOGICAL DYNAMICS: Furstenberg Families and Ellis Actions
Ethan Akin

A SHORT INTRODUCTION TO INTUITIONISTIC LOGIC
Grigori Mints

TOPICS IN NUMBER THEORY
J. S. Chahal

VARIATIONS ON A THEME OF EULER: Quadratic Forms, Elliptic Curves, and Hopf Maps
Takashi Ono

A Short Introduction to Intuitionistic Logic

Grigori Mints
Stanford University
Stanford, California

Kluwer Academic / Plenum Publishers
New York, Boston, Dordrecht, London, Moscow

Library of Congress Cataloging-in-Publication Data

On file

ISBN 0-306-46394-6

233 Spring Street, New York, New York 10013

http://www.wkap.nl/

10 9 8 7 6 5 4 3 2 1

A C.I.P. record for this book is available from the Library of Congress

Printed in the United States of America

Preface

This book is an introduction to intuitionistic logic that stresses the subject's connections with computer science. To make the material more accessible, basic techniques are presented first for propositional logic; Part II contains extensions to predicate logic. This material provides a safe background for reading research literature in logic and computer science as well as such advanced monographs as [26, 27], [25],[13],[24]. Readers are assumed to be familiar with basic notions of first order logic presented for example in introductory chapters of such books as [12], [28], [5]. Subsections on algebraic and topological semantics require some initial information in algebra and topology. One device for making this book short was inventing new (or at least modified) proofs of several well-known theorems.

For historic perspective and credits readers may consult Notes at the end of chapters in [26, 27] or sections in [25]. Suggestions for further reading are found in our Introduction.

This text developed from material for several courses taught at Stanford University in 1992–99. Special thanks are due to N. Bjørner, who took some of these courses, asked profound questions, and typed course notes in LaTeX, and to A. Everett, who carefully checked a preliminary version in English.

Stanford, December 1999 Grigori Mints

Contents

Introduction

Intuitionistic logic is studied here as part of familiar classical logic which allows an effective interpretation and mechanical extraction of programs from proofs. The possibility of retaining classical patterns of reasoning is emphasized, while warning signs of differences (beginning with the invalidity of the excluded middle $p \vee \neg p$) are presented only when strictly necessary.

Traditional, or classical logic, is traceable to Aristotle. It took its present form at the end of nineteenth and beginning of twentieth centuries. Intuitionistic logic, the subject of this book has its source in the work of L. E. J. Brouwer.

The aspect that turned out to be the most important was his requirement of effective existence: To claim

$$\exists x A,$$

we should specify an object t, and justify

$$A[x/t].$$

Similarly to claim a disjunction:

$$A_1 \vee A_2,$$

we should indicate which of the disjuncts is true; that is, pick an $i \in \{1, 2\}$, and justify

$$A_i.$$

Brouwer stressed that classical logic in general does not satisfy these conditions.

He drew attention to the *law of the excluded middle* or *excluded third* :

$$A \vee \neg A$$

It is not clear how to discover in general which of the disjuncts is true. Brouwer gave very serious arguments for the falsity of this law under his effective interpretation. He and his followers suggested restricting methods of reasoning to those that are safe in this respect.

Our approach to this controversy is pragmatic: It turned out that the logical means recognized as correct by Brouwer's school allow *construction of correct programs and correctness proofs* for these.

Brouwer's philosophic justification of his foundational conception referred to mathematical intuition. He called his philosophical system *intuitionism*; formal systems using means correct from the intuitionistic point of view are also called *intuitionistic* formal systems.

The systems of first-order intuitionistic logic and arithmetic have a very simple characterization: these are obtained from suitable traditional or classical systems by removing the law of the excluded middle or equivalent principles, such as the rule of double negation.

Our book is organized as follows:

The first three Chapters of Part I explain notation for natural deduction to be used in the book, introduce the reader acquainted with other kinds of formal systems to this apparatus, and present some basic information to be used later. Intuitionistic propositional system NJp is introduced here.

Chapter 3 describes a translation of classical into intuitionistic logic, which is an identity from the classical point of view. This translation allows us to consider classical logic as a subset (consisting of negative formulas) of intuitionistic logic.

Chapter 5 introduces a programming interpretation of intuitionistic propositional logic, which is a significant part of its interest and a source of many applications and extensions in various directions. This construction demonstrates that programs (simply typed λ-terms) are actually the same as natural deductions, which justifies such expressions and slogans as the Curry–Howard isomorphism and:

PROOFS = PROGRAMS

A simple normalization proof (going back to Turing (see [7]) is presented and such standard consequences as disjunction property, Harrop's theorem for disjunction, underivability of the excluded middle are drawn in Section 5.4.. More sophisticated consequences that distinguish a class of formulas important in applications to category theory and having unique derivations are drawn in Chapter 6.

Chapter 7 introduces propositional Kripke models and their most important special cases. Chapter 8 presents a construction of the canonical model and a completeness proof (see [6]) establishing cut elimination and a finite model property for a multiple-succedent L-style formalization LJpm of intuitionistic propositional logic. This proof reintroduces the theme of proof-search by saturating a given goal with respect to all available rules that appeared in Section 2.6..

Chapter 9 contains a completeness proof for topological models on the segment [0,1] of reals accessible to readers with a very restricted background in general topology.

Chapter 10 uses an example of intuitionistic propositional logic to present basic ideas of proof-search methods that proceed from a goal to simpler subgoals. Such methods are employed in many working systems for proof checking and automated deduction, and these methods form starting points of more sophisticated algorithms. The resulting completeness proof exemplifies yet another proof of the normal form theorem (completeness of cut-free formulation) dating from Gödel's proof of his completeness theorem and stresses the close connection between proof-search strategies and Kripke semantics.

Chapter 11 introduces a new technique, permutation of inferences in a cut-free derivations, which is used to restrict derivations further and leads to one-succedent version LJp of intuitionistic propositional logic, introduced by Gentzen [8]. The LJp is applied to give an easy construction that proves the Craig interpolation theorem and Beth definability theorem.

The second part of the book presents primarily extensions of these methods and results to predicate logic and spells out necessary changes in formulations and proofs. One of the most prominent differences is the definition of the proof-search procedure. It follows Kripke [14] in adding a new rule of *transfer*. In fact it is possible to avoid this rule in the same way as in the propositional case, but this creates unnecessary redundancy and complicates proof-search strategy too much.

Topics for Further Reading

Permutative conversions that remove remote cuts when introduction and elimination inferences are separated by a segment consisting of ∨-inferences are not included in Chapter 5. As a result some normal derivations do not have the subformula property. However a sufficient number of the basic applications of full normalization is still available, and including permutative conversion would direct a beginner's attention in the wrong direction. A full normalization proof is presented in many places, for example [19], [25]. Subformula property becomes available later for L-style systems (with rules for an introduction to antecedent and succedent).

Another extension omitted here is a coherence theorem for Cartesian closed categories. An easy proof for the →-fragment in Chapter 8 provides enough information and intuition to go if necessary through technical details of the proof for $\{\&, \rightarrow\}$-language in [16] (reproduced in [25]).

One of the basic results of proof theory is a syntactic normalization (cut elimination) theorem for an L-style system LJ: Every derivation in LJ+(cut rule) can be reduced to a cut-free derivation by a finite number of local reductions similar to normalization steps for NJ. This result is not included here, since it duplicates many features of the normalization proof for NJ (Chapter 5) and there are many good presentations of cut elimination, such as [8],[13],[24],[25]. Moreover a *normal form* result claiming that derivability in LJm+cut implies derivability in LJm (without connection between derivations with and without cut) is proved in Chapters 11 and 15.

Another important result in the same direction is *strong normalization* for NJ-deductions (or deductive terms): Every sequence of reductions beginning with given deduction terminates in a normal form, and the normal form is unique (Church-Rosser property). Readers are referred to [25] and [26, 27] for proofs. Both theory and practice of automated deduction for classical predicate logic depend on Skolemization and Herbrand theorem. These results extend to intuitionistic logic in only a restricted (but still useful) ways; see [15] and [21].

The next step in automated deduction after bottom-up (goal-to-subgoal) proof-search is resolution; see [4] for classical logic and [17] and [25] for intuitionistic logic. Recursive undecidability of intuitionistic predicate logic follows from the negative interpretation of classical logic (Theorem 16, Section 13.2.).

A decision procedure for intuitionistic propositional logic is described in Section 10.3.; in fact this logic is PSPACE-complete [22].

Algebraic and topological completeness results for propositional logic (Section 8.4., Chapter 9) extend to predicate logic [20].

Forty to fifty years ago quite a different presentation of a logical system (than natural deduction) was standard: A series of axioms and one inference rule, *modus ponens* (and perhaps one or two rules for quantifiers). Such Hilbert-type axiomatizations are given in [9], [8], [25] as well as many other places.

Underivability results for intuitionistic logic are mentioned mainly to warn readers at the very beginning and to illustrate the use of model-theoretic methods later. More detailed information are found in [13], [19], [25].

Most automated deduction programs for intuitionistic logic provide derivations in such systems as LJm or even LJm*. To extract programs from such derivations, it is possible to use term assignments more ot less mimicking Kripke translation χ in Chapter 16. Optimization of arising redundancies along the lines of Section 11.3. is discussed in [3], [23]. These techniques have their origin in the translation of cut-free L-style derivations into normal natural deductions described in [19].

Part I

Intuitionistic Propositional Logic

Chapter 1

Preliminaries

We use such standard abbreviations as iff for "if and only if", IH for Induction Hypothesis (in proofs by mathematical induction), and so on; the symbol $\equiv$ stands for syntactic identity of expressions, and $:=$ denotes equality by definition. The symbol $\dashv$ indicates the end a proof. We assume some basic knowledge of the properties of free and bound variables in formulas of predicate logic; in the treatment of substitution, we follow standard conventions (see for example [25]). Expressions that differ only in the name of bound variables are regarded as identical. In the definition of substituting the expression e' for the variable x in expression e, either one requires no variable free in e' to become bound by an operator like $\forall x$, $\exists x$, λx in e (that is, e' is free for x in e, or there are no clashes of variables), or the substitution operation is taken to involve a systematic renaming operation for bound variables, thereby avoiding clashes. Having stated that we are interested only in expressions modulo renaming bound variables, we can without loss of generality assume that substitution is always possible. The abbreviation FV is used for the set of free variables of an expression.

It is often convenient to work with *finite multisets*, that is, finite sequences modulo the ordering: Permutation $\alpha_{i_1}, \ldots, \alpha_{i_n}$ is identified with $\alpha_1, \ldots, \alpha_n$, but the number of occurrences of each α_i is important. Multisets of formulas are denoted by $\Gamma, \Gamma_1, \Delta, \Delta_1, \ldots$. The notation α, Γ is used for the result of adding (one more occurrence of) α to the multiset Γ. The notation Γ, Σ means the multiset union of Γ and Σ: Occurrences of the same formula are not contracted, for example, $\{\alpha, \alpha\}, \{\alpha, \alpha, \alpha\}$ is $\{\alpha, \alpha, \alpha, \alpha, \alpha\}$.

The $[\Gamma, \Delta]$ represents the multiset union of Γ, Δ and a possible identification of some formulas in Δ with identical formulas in Γ. For example:

$$[\{\alpha, \alpha, \beta, \beta\}, \{\alpha, \alpha, \alpha, \gamma, \gamma\}]$$

can be any of:

$$\{\alpha, \alpha, \alpha, \alpha, \alpha, \beta, \beta, \gamma, \gamma\}, \{\alpha, \alpha, \alpha, \alpha, \beta, \beta, \gamma, \gamma\}, \{\alpha, \alpha, \alpha, \beta, \beta, \gamma, \gamma\},$$

but not $\{\alpha, \alpha, \beta, \beta, \gamma, \gamma\}$.

Chapter 2

Natural Deduction for Propositional Logic

Formalization of intuitionistic logic was obtained by dropping some axioms of *classical propositional logic* or *classical propositional calculus* abbreviated CPC . Readers are assumed to know basic properties of this latter system. In the following we recall some facts about CPC and fix notation.

2.1. Syntax

Formulas of propositional logic are constructed in the standard way from *propositional variables* or *propositional letters* and a constant $\perp$ by means of *logical connectives* $\vee, \&, \rightarrow$. Symbols $p, q, r, s, p_1, q_1, r_1$ and so on, represent propositional variables. Symbols $\alpha, \beta, \gamma, \ldots, \alpha_1, \beta_1, \gamma_1, \ldots$ represent formulas.

Other connectives and the constant $\top$ are treated as abbreviations:

$$\neg\alpha := (\alpha \rightarrow \perp), \quad \top = \neg\perp, \quad p \leftrightarrow q := ((p \rightarrow q)\&(q \rightarrow p))$$

EXAMPLE 2.1. $p, q, (p\&q), (p \vee (p \rightarrow q)), (((p \rightarrow \neg q) \rightarrow p) \leftrightarrow p)$ are formulas.

We often drop outermost parentheses as well as parentheses dividing terms in a conjunction or a disjunction.

Let us describe our basic system NJp of intuitionistic propositional logic. The additional rule $\perp_c$ of double negation leads to a system NKp that is complete for deriving propositional tautologies, that is, classically valid formulas. It is convenient to work with *sequents*:

$$\alpha_1, \ldots, \alpha_n \Rightarrow \alpha$$

read "*Assumptions* $\alpha_1, \ldots, \alpha_n$ imply α" or $\alpha_1\& \ldots \&\alpha_n \rightarrow \alpha$. The formula α is called the *succedent* of the sequent. The part $\alpha_1, \ldots, \alpha_n$ to the left of $\Rightarrow$ is called the *antecedent*, and it is treated as a *multiset* (see Section 1).

2.2. Intuitionistic Propositional System NJp

Axioms:

$$\alpha \Rightarrow \alpha$$

Inference rules (I, E stand for introduction, elimination):

$$\frac{\Gamma \Rightarrow \alpha \quad \Delta \Rightarrow \beta}{[\Gamma, \Delta] \Rightarrow \alpha \& \beta} \& I \qquad \frac{\Gamma \Rightarrow \alpha \& \beta}{\Gamma \Rightarrow \alpha} \& E \qquad \frac{\Gamma \Rightarrow \alpha \& \beta}{\Gamma \Rightarrow \beta} \& E$$

$$\frac{\Gamma \Rightarrow (\alpha \to \beta) \quad \Delta \Rightarrow \alpha}{[\Gamma, \Delta] \Rightarrow \beta} \to E \qquad \frac{\alpha^0, \Gamma \Rightarrow \beta}{\Gamma \Rightarrow (\alpha \to \beta)} \to I$$

$$\frac{\Gamma \Rightarrow \alpha}{\Gamma \Rightarrow (\alpha \vee \beta)} \vee I \qquad \frac{\Gamma \Rightarrow \beta}{\Gamma \Rightarrow (\alpha \vee \beta)} \vee I$$

$$\frac{\Gamma \Rightarrow \alpha \vee \beta \quad \alpha^0, \Delta \Rightarrow \phi \quad \beta^0, \Sigma \Rightarrow \phi}{[\Gamma, \Delta, \Sigma] \Rightarrow \phi} \vee E$$

$$\frac{\Gamma \Rightarrow \bot}{\Gamma \Rightarrow \theta} \bot_i$$

The superscript ^{0} in the rules $\to I, \vee E$ means that the corresponding assumptions may be absent. The formula θ in the rule $\bot_i$ is atomic (but this restriction is removed in Section 2.5.).

Two further rules are derivable in NJp:

$$\frac{\alpha, \alpha, \Gamma \Rightarrow \phi}{\alpha, \Gamma \Rightarrow \phi} \textit{contr} \qquad \frac{\Gamma \Rightarrow \phi}{\alpha, \Gamma \Rightarrow \phi} \textit{weak}$$

2.3. Classical Propositional System NKp

Classical Propositional Calculus is obtained by adding the classical negation rule $\bot_c$ for proofs *ad asurdum* to the system NJp:

$$\frac{\neg\alpha, \Gamma \Rightarrow \bot}{\Gamma \Rightarrow \alpha} \bot_c$$

A *natural deduction* or a *proof* in the system NJp or NKp is defined in a standard way: It is a tree beginning with axioms and proceeding by the inference rules of the system. A sequents is *deducible* or *provable* if it is a last sequent of a deduction. A formula α is *deducible* or *provable* if the sequent $\Rightarrow \alpha$ is provable.

Notation $d : \Gamma \Rightarrow \alpha$ indicates that d is a natural deduction of $\Gamma \Rightarrow \alpha$, and $\Gamma \vdash \alpha$ means that the sequent $\Gamma \Rightarrow \alpha$ is derivable in NJp.

Axiom $\alpha \Rightarrow \alpha$ introduces *assumption* α. We sometimes treat weakening of axioms, that is sequents $\Gamma, \alpha, \Delta \Rightarrow \alpha$ as axioms. Applications of inference rules (or *inferences*) transform goal formulas written to the right of $\Rightarrow$ and leave assumptions intact except for $\to I$ and $\vee E$-inferences, which *discharge* the assumption α ($\vee E$ discharges also assumption β). Every connective $\odot$ has two rules: An introduction rule $\odot I$ for proving a formula beginning with the connective and an elimination rule $\odot E$ for deriving consequences from proved formula beginning with this connective.

Unless stated otherwise, we are interested in derivability in the system NJp. The most important exception is Section 2.9., and after that most of our deductions do not use the rule $\bot_c$.

EXAMPLE 2.2. The formula $p \rightarrow (q \rightarrow p)$ is derived in NJp by two $\rightarrow I$-inferences from the axiom $p \Rightarrow p$:

$$\dfrac{\dfrac{p \Rightarrow p}{p \Rightarrow q \rightarrow p}\rightarrow I}{\Rightarrow p \rightarrow (q \rightarrow p)}\rightarrow I$$

EXAMPLE 2.3. The formula $(p \rightarrow (q \rightarrow r)) \rightarrow ((p \rightarrow q) \rightarrow (p \rightarrow r))$ is derived as follows: Assume premises $p \rightarrow (q \rightarrow r), p \rightarrow q, p$ of the implication and use $\rightarrow E$ to infer successively $q \rightarrow r, q$ and r. Then use $\rightarrow I$ three times to recover the whole formula:

$$\dfrac{\dfrac{\dfrac{\dfrac{\dfrac{p\rightarrow(q\rightarrow r)\Rightarrow p\rightarrow(q\rightarrow r)\quad p\Rightarrow p}{p\rightarrow(q\rightarrow r),p\Rightarrow q\rightarrow r}\rightarrow E\quad \dfrac{p\rightarrow q\Rightarrow p\rightarrow q\quad p\Rightarrow p}{p\rightarrow q,p\Rightarrow q}\rightarrow E}{p\rightarrow(q\rightarrow r),p\rightarrow q,p\Rightarrow r}\rightarrow E}{p\rightarrow(q\rightarrow r),p\rightarrow q\Rightarrow p\rightarrow r}\rightarrow I}{p\rightarrow(q\rightarrow r)\Rightarrow(p\rightarrow q)\rightarrow(p\rightarrow r)}\rightarrow I}{d:\Rightarrow (p\rightarrow(q\rightarrow r))\rightarrow((p\rightarrow q)\rightarrow(p\rightarrow r))}\rightarrow I$$

We do not attach rule annotations like $\rightarrow E, \rightarrow I$ in the deductions that follow.

2.4. Abbreviated Notation for Natural Deductions

For a natural deduction d, let d^- be the result of deleting all the assumptions and $\Rightarrow$ (as well as repetitions of the sequents) from d. For example the previous deduction d becomes

$$\dfrac{\dfrac{\dfrac{\dfrac{\dfrac{p\rightarrow(q\rightarrow r)\quad p}{q\rightarrow r}\quad \dfrac{p\rightarrow q\quad p}{q}}{r}}{p\rightarrow r}}{(p\rightarrow q)\rightarrow(p\rightarrow r)}}{d^- :\ (p\rightarrow(q\rightarrow r))\rightarrow((p\rightarrow q)\rightarrow(p\rightarrow r))}$$

The derivation d can be recovered from d^- (and some information on assumptions that can be disregarded at this stage).

EXAMPLE 2.4. To derive $(p\&q) \to p$, assume the premise and apply $\&E$ followed by $\to I$:

$$\dfrac{\dfrac{p\&q}{p}}{(p\&q) \to p}$$

EXAMPLE 2.5. To derive $p \to (q \to (p\&q))$, assume both premises and apply $\&I$:

$$\dfrac{\dfrac{\dfrac{p \quad q}{p\&q}}{q \to (p\&q)}}{p \to (q \to (p\&q))}$$

EXAMPLE 2.6. For $p \to (p \vee q)$ use $\vee I$:

$$\dfrac{\dfrac{p}{p \vee q}}{p \to (p \vee q)}$$

EXAMPLE 2.7. For $(p \to r) \to ((q \to r) \to ((p \vee q) \to r))$, assume all premises and use $\vee E$ combined with deductions of r from premises. We show a nonabbreviated deduction here to demonstrate handling assumptions in $\vee E$-inference:

$$\dfrac{\dfrac{\dfrac{\dfrac{p \vee q \Rightarrow p \vee q \qquad \dfrac{p \to r \Rightarrow p \to r \quad p \Rightarrow p}{p, p \to r \Rightarrow r} \qquad \dfrac{q \to r \Rightarrow q \to r \quad q \Rightarrow q}{q, q \to r \Rightarrow r}}{p \to r, q \to r, p \vee q \Rightarrow r}}{p \to r, q \to r \Rightarrow (p \vee q) \to r}}{p \to r \Rightarrow (q \to r) \to ((p \vee q) \to r)}}{\Rightarrow (p \to r) \to ((q \to r) \to ((p \vee q) \to r))}$$

The next two Examples illustrate treating negation $\neg\alpha$ as an abbreviation for the implication $\alpha \to \bot$.

EXAMPLE 2.8. For $\neg p \to (p \to q)$, apply $\to E$ (and recall that $\neg p$ is an abbreviation for $p \to \bot$):

$$\dfrac{\dfrac{\dfrac{\dfrac{\neg p \quad p}{\bot}}{q}}{p \to q}}{\neg p \to (p \to q)}$$

EXAMPLE 2.9. For $(p \to q) \to ((p \to \neg q) \to \neg p)$, apply $\to I$:

$$\dfrac{\dfrac{\dfrac{\dfrac{\dfrac{p \to \neg q \quad p}{\neg q} \quad \dfrac{p \to q \quad p}{q}}{\bot}}{\neg p}}{(p \to \neg q) \to \neg p}}{(p \to q) \to ((p \to \neg q) \to \neg p)}$$

EXAMPLE 2.10. Here is a derivation of $\neg\neg(p \lor \neg p)$:

$$\dfrac{\dfrac{\neg(p \lor \neg p) \qquad \dfrac{\dfrac{\dfrac{\neg(p \lor \neg p) \quad \dfrac{p}{p \lor \neg p}}{\bot}}{\neg p}}{p \lor \neg p}}{\bot}}{\neg\neg(p \lor \neg p)}$$

2.5. Derivable Rules

It is often convenient to treat a series of inference rules as one rule.

DEFINITION 2.1. *A* deduction *of a sequent S* from *sequents $S_1, \dots, S_n$ is a tree beginning with axioms or sequents $S_1, \dots, S_n$ and proceeding by inference rules.*

A rule

$$\frac{S_1, \dots, S_n}{S}$$

is derivable *if there is a deduction of S from $S_1, \dots, S_n$.*

EXAMPLE 2.11. The cut rule:

$$\frac{\Gamma \Rightarrow \alpha \quad \alpha, \Gamma \Rightarrow \beta}{\Gamma \Rightarrow \beta}\ cut$$

is derived as follows:

$$\dfrac{\dfrac{\alpha, \Gamma \Rightarrow \beta}{\Gamma \Rightarrow \alpha \to \beta} \quad \Gamma \Rightarrow \alpha}{\Gamma \Rightarrow \beta}$$

Weakening and contraction are derivable as follows:

$$\dfrac{\dfrac{\Gamma \Rightarrow \alpha}{\Gamma \Rightarrow \beta \to \alpha} \quad \beta \Rightarrow \beta}{\beta, \Gamma \Rightarrow \alpha} \qquad \dfrac{\dfrac{\dfrac{\beta, \beta, \Gamma \to \alpha}{\Gamma \Rightarrow \beta \to (\beta \to \alpha)} \quad \beta \Rightarrow \beta}{\beta, \Gamma \Rightarrow \beta \to \alpha} \qquad \beta \Rightarrow \beta}{\beta, \Gamma \Rightarrow \alpha}$$

LEMMA 2.1. *The rule* $\bot_i$ *with arbitrary* α*:*

$$\frac{\Gamma \Rightarrow \bot}{\Gamma \Rightarrow \alpha}$$

is derivable.

Proof. We use induction on α. Induction base (α is atomic) is the rule of NJp. Induction step (composite $\alpha \equiv \phi \odot \psi$ for $\odot \equiv \&, \vee, \rightarrow$) uses the introduction rule for $\odot$, for example:

$$\dfrac{\text{by IH } \dfrac{\Gamma \Rightarrow \bot}{\Gamma \Rightarrow \phi} \quad \text{by IH } \dfrac{\Gamma \Rightarrow \bot}{\Gamma \Rightarrow \psi}}{\Gamma \Rightarrow \phi \& \psi} \; \&I$$

⊣

LEMMA 2.2. *The rule*

$$\frac{\Gamma \Rightarrow \alpha_1, \ldots, \Gamma \Rightarrow \alpha_n}{\Gamma \Rightarrow \beta} \; (R)$$

is derivable iff the sequent $\Gamma \Rightarrow \alpha_1 \rightarrow (\alpha_2 \rightarrow \ldots \rightarrow (\alpha_n \rightarrow \beta) \ldots) \equiv \Gamma \Rightarrow \alpha_R$ *is derivable.*

Proof.(R) is obtained from $\Gamma \Rightarrow \alpha_R$ by a series of $\rightarrow E$. Vice versa, $\Gamma \Rightarrow \alpha_R$ is obtained from $\Gamma, \alpha_1, \ldots, \alpha_i, \ldots, \alpha_n \Rightarrow \alpha_i$, $(i = 1, \ldots, n)$ by (R) and a series of $\rightarrow I$. ⊣

EXAMPLE 2.12. An instance of the rule in the previous Lemma:

$$\frac{\Gamma \Rightarrow \alpha_1 \rightarrow \alpha_2 \ldots \Gamma \Rightarrow \alpha_{n-1} \rightarrow \alpha_n}{\Gamma \Rightarrow \alpha_1 \rightarrow \alpha_n} \; Trans$$

EXAMPLE 2.13. Relaxed versions of two-premise rules. For every derivable rule:

$$\frac{\Gamma \Rightarrow \alpha \quad \Gamma \Rightarrow \beta}{\Gamma \rightarrow \phi} \; R$$

the rule

$$\frac{\Gamma \Rightarrow \alpha \quad \Delta \Rightarrow \beta}{[\Gamma, \Delta] \rightarrow \phi}$$

is derivable too, since the sequent $\alpha \rightarrow (\beta \rightarrow \phi) \Rightarrow \alpha \rightarrow ((\sigma \rightarrow \beta) \rightarrow (\sigma \rightarrow \phi))$ is derivable. Hence it is possible to assume the assumptions Γ are the same in all premises of the rule.

2.6. Direct Chaining and Analysis into Subgoals

The problem of deriving given sequent (goal) S can often be reduced to deriving simpler sequents (subgoals), say, S_1, S_2, if the rule

$$\frac{S_1 \quad S_2}{S}$$

is derivable. Let us list some of these rules.

Rules for analysis into subgoals

LEMMA 2.3. *The following rules are derivable in NJp:* $\rightarrow I, \& I, \vee I$,

$$\frac{\alpha, \Gamma \Rightarrow \beta \quad \beta, \Gamma \Rightarrow \alpha}{\Gamma \Rightarrow \alpha \leftrightarrow \beta} \leftrightarrow I \qquad \frac{\alpha, \Gamma \Rightarrow \phi \quad \beta, \Gamma \Rightarrow \phi}{\alpha \vee \beta, \Gamma \Rightarrow \phi} \vee \Rightarrow \qquad \frac{\alpha^0, \Gamma \Rightarrow \bot}{\Gamma \Rightarrow \neg\alpha} \neg I$$

Proof. $\leftrightarrow I$: expand abbreviation $\alpha \leftrightarrow \beta$ into $(\alpha \rightarrow \beta)\&(\beta \rightarrow \alpha)$ and use $\rightarrow I, \& I$.

$\vee \Rightarrow$: use $\vee E$ with the first premise $\alpha \vee \beta \Rightarrow \alpha \vee \beta$. ⊣

Rules for direct chaining

Sometimes applying elimination (and very similar) rules to assumptions quickly produces the succedent of the sequent. Some of these rules follow.

LEMMA 2.4. *The following rules are derivable in NJp:* $\rightarrow E, \& E, \neg E, \bot_i$, $\vee \Rightarrow$ *(see Lemma 2.6.), Trans (Example 2.5.),*

$$\frac{\Gamma \Rightarrow \alpha \leftrightarrow \beta}{\Gamma \Rightarrow \alpha \rightarrow \beta} \leftrightarrow E \qquad \frac{\Gamma \Rightarrow \alpha \leftrightarrow \beta}{\Gamma \Rightarrow \beta \rightarrow \alpha} \leftrightarrow E \qquad \frac{\Gamma \rightarrow \bot}{\Gamma \Rightarrow \beta} \bot_i \qquad \frac{\Gamma \Rightarrow \neg\alpha \quad \Delta \Rightarrow \alpha}{[\Gamma, \Delta] \Rightarrow \beta} \neg E$$

Proof. $\leftrightarrow E$: expand $\alpha \leftrightarrow \beta$ and use $\& E$. ⊣

DEFINITION 2.2. *A deduction using only rules mentioned in Lemma 2.6., cut and structural rules is called* direct chaining.

Note. A good heuristic for deducing $\Gamma \Rightarrow \alpha$ by direct chaining is to take Γ as the initial set of data and saturate it by adding conclusions of all the rules except $\vee \Rightarrow$, mentioned in Lemma 2.6. plus cut (Example 2.5.), restricting applications of $\bot_i$ to subformulas of Γ, α, producing say $\Gamma_1 \equiv \Gamma^+$. Stop if α is obtained, otherwise apply $\vee \Rightarrow$ bottom-up for each formula $\alpha \vee \beta \in \Gamma$, that is, form $(\alpha, \Gamma_1)^+$ and $(\beta, \Gamma_1)^+$. Now add all formulas $\gamma \in (\alpha, \Gamma_1)^+ \cup (\beta, \Gamma_1)^+$ to Γ_1 forming Γ_2. Iterate the process till it stops.

EXAMPLE 2.14. $p_0, p_0 \rightarrow p_1, \ldots, p_{n-1} \rightarrow p_n \Rightarrow p_n$ is obtained by $n \rightarrow E$-inferences.

EXERCISE 2.1. Derive $p_0, p_0 \rightarrow p_1 \& p_2, p_1 \rightarrow p_3, p_2 \rightarrow p_4, p_3 \rightarrow (p_4 \rightarrow p_5) \Rightarrow p_5$ by direct chaining.

ADC method of establishing deducibility

The approach opposite to direct chaining is often useful for reducing the goal formula. In Examples 2.3. and 2.3., when the goal was an implication $\phi \equiv \alpha_1 \to (\ldots \to (\alpha_n \to \psi)\ldots)$, it was reduced to a sequent $\alpha_1, \ldots, \alpha_n \Rightarrow \psi$; that is, this sequent was derived, then $\to \phi$ was deduced from it by $n \to I$-inferences. In other words, the *goal* ϕ was successively reduced to *subgoals* $\phi_1 \equiv \alpha_2 \Rightarrow (\ldots \to (\alpha_n \to \psi)\ldots)$, ..., $\phi_n \equiv \alpha_1, \ldots, \alpha_n \Rightarrow \psi$. A natural deduction of $\to \phi$ was constructed *from the bottom up*:

$$\cfrac{\cfrac{\phi_n \equiv \alpha_1, \ldots, \alpha_n \Rightarrow \psi}{\vdots}}{\cfrac{\phi_1 \equiv \alpha_1 \Rightarrow \alpha_2 \to (\ldots \to (\alpha_n \to \psi)\ldots)}{\phi \equiv \alpha_1 \to (\ldots \to (\alpha_n \to \psi)\ldots)}}$$

One of the most straightforward methods of establishing deducibility of a sequent $\Gamma \Rightarrow \alpha$ consists in its analysis into subgoals $\Gamma_1 \Rightarrow \alpha_1, \ldots, \Gamma_n \Rightarrow \alpha_n$ using Lemma 2.6., and establishing each subgoal by direct chaining. We say that a sequent $\Delta \Rightarrow \alpha \vee \beta$ is established when one of $\Delta \Rightarrow \alpha$, $\Delta \Rightarrow \beta$ is established. The combination of Analysis and Direct Chaining described above will be called ADC-*method* or simply ADC. It is not complete: some valid formulas are not deducible by ADC.

PROBLEM 2.1. Estimate the complexity of ADC and find a suitable subclass for which it is complete.

2.7. Heuristics for Natural Deduction

Recall that $\Gamma \vdash \phi$ means that the sequent $\Gamma \Rightarrow \phi$ is derivable in NJp.

LEMMA 2.5. *Every sequent derivable in NKp (and hence in NJp) is a tautology according to classical truth tables (see Chapter 2.9.).*

Proof. For a given truth value assignment v define

$$v(\alpha_1, \ldots, \alpha_n \Rightarrow \alpha) := v(\&_{i \leq n} \alpha_i \to \alpha) := \max(1 - \min(v(\alpha_1), \ldots, v(\alpha_n)), v(\alpha))$$

Check that every rule preserves truth (that is, value 1) under every assignment v. For example, for the rule $\to E$ assume that $v(\&\Gamma \to (\alpha \to \beta)) = 1$ and $v(\&\Gamma \to \alpha) = 1$. We must prove $v(\&\Gamma \to \beta) = 1$. If $v(\&\Gamma) = 0$, we are done. Otherwise $v(\alpha \to \beta) = 1$ and $v(\alpha) = 1$; hence $v(\beta) = 1$: See the truth table for $\to$. ⊣

The converse of the previous Lemma is Theorem 2.9.2below. As stated in the Introduction, and proved in Section 5.4., some tautologies including $\neg\neg\alpha \to \alpha$ are not derivable in NJp.

LEMMA 2.6. $\Gamma \vdash \alpha \to \beta$ *iff* $\Gamma, \alpha \vdash \beta$

One direction is $\to E$, the other direction is $\to I$:

$$\frac{\Gamma \Rightarrow \alpha \to \beta \quad \alpha \Rightarrow \alpha}{\Gamma, \alpha \Rightarrow \beta} \qquad \frac{\Gamma, \alpha \Rightarrow \beta}{\Gamma \Rightarrow \alpha \to \beta}$$

$\dashv$

EXAMPLE 2.15. Deduce $\alpha \to q \Rightarrow q$ where $\alpha \equiv ((p \to q) \to p) \to p$. By Lemma 2.7. it is sufficient to deduce $\alpha \to q, (p \to q) \to p \Rightarrow p$, which is reduced to $\alpha \to q, p \Rightarrow q$, then to $p \Rightarrow \alpha$, then to $p, (p \to q) \to p \Rightarrow p$, which is a (weakening of an) axiom.

Let us write the deduction in an abbreviated form:

$$\dfrac{\alpha \to q \qquad \dfrac{\dfrac{(p \to q) \to p \qquad \dfrac{\dfrac{\alpha \to q \quad \dfrac{p}{\alpha}}{q}}{p \to q}}{p}}{\alpha}}{q}$$

Heuristics for Disjunction

LEMMA 2.7. *$\alpha \vee \beta, \Gamma \vdash \phi$ iff $\alpha, \Gamma \vdash \phi$ and $\beta, \Gamma \Rightarrow \phi$.*

Proof. $\alpha \vee \beta, \Gamma \vdash \phi$ implies $\alpha, \Gamma \vdash \phi$ by $\alpha \vdash \alpha \vee \beta$. On the other hand, $\alpha, \Gamma \vdash \phi$ and $\beta, \Gamma \Rightarrow \phi$ imply $\alpha \vee \beta, \Gamma \vdash \phi$ by $\vee E$. $\dashv$

Heuristics for Negation

Recall that $\neg\alpha \equiv (\alpha \to \bot)$. We list some provable properties of negation to be used later.

LEMMA 2.8. *(a) $\Gamma \vdash \neg\alpha$ iff $\Gamma, \alpha \vdash \bot$*

(b) double negation introduction and elimination:

(b1) $\vdash \alpha \to \neg\neg\alpha$

(b2) $\vdash \neg\neg\neg\alpha \leftrightarrow \neg\alpha$

(b3) $\vdash \neg\neg\bot \leftrightarrow \bot$

(c) de Morgan's laws valid intuitionistically:

$\vdash \neg(\alpha \vee \beta) \leftrightarrow (\neg\alpha \& \neg\beta)$

$\vdash \neg(\alpha \& \beta) \leftrightarrow (\alpha \to \neg\beta)$

$\vdash \neg(\alpha \to \beta) \leftrightarrow (\neg\neg\alpha \& \neg\beta)$

(d) $\Gamma \vdash \bot$ iff $\Gamma \vdash \alpha$, $\Gamma \vdash \neg\alpha$ for some α

Proof.(a) Expand $\neg\alpha$ into $\alpha \to \bot$ and use Lemma 2.7..

(b1) Use ADC. (b2) Use ADC and (b1). (b3) $\neg\neg\bot \equiv (\bot \to \bot) \to \bot$. Detach derivable $\bot \to \bot$.

(c) Use ADC and the introduction rule for the negated connective. For example, $\neg(\alpha \vee \beta) \to (\neg\alpha \& \neg\beta)$ is reduced by analysis to $\neg(\alpha \vee \beta), \alpha \vdash \bot$ and $\neg(\alpha \vee \beta), \beta \vdash \bot$; it remains to apply $\alpha \vdash \alpha \vee \beta$.

(d) Use $\neg E, \bot_i$. ⊣

LEMMA 2.9.

$$\alpha \to \beta, \neg\alpha \to \beta \Rightarrow \beta \quad \textit{is derivable in NKp} \tag{2.1}$$

$$\alpha \to \beta, \neg\alpha \to \beta \Rightarrow \neg\neg\beta \quad \textit{is derivable in NJp} \tag{2.2}$$

Proof. First prove in NJp that:

$$\neg\beta, \alpha \to \beta, \neg\alpha \to \beta \vdash \bot \tag{2.3}$$

by direct chaining via $\neg\beta, \alpha \to \beta \vdash \neg\alpha$. Now use (2.3) as a premise of the $\to I$-rule in NJp to obtain (2.2), and as a premise of the $\bot_c$-rule in NKp to derive (2.1). ⊣

Note. This is the first time we used $\bot_c$-rule.

In Chapter 3 we prove that $\Gamma \vdash \bot$ in NJp iff $\Gamma \Rightarrow \bot$ is a tautology (Glivenko's theorem).

2.8. Replacement of Equivalents

LEMMA 2.10. *(a) for every connective $\odot \in \{\&, \vee, \to\}$:*

$$\vdash (\alpha \leftrightarrow \beta) \to ((\alpha \odot \gamma) \leftrightarrow (\beta \odot \gamma))$$

$$\vdash (\alpha \leftrightarrow \beta) \to ((\gamma \odot \alpha) \leftrightarrow (\gamma \odot \beta))$$

(b) In general for every formula γ and every propositional variable p:

$$\vdash (\alpha \leftrightarrow \beta) \to (\gamma[p/\alpha] \leftrightarrow \gamma[p/\beta]),$$

where $\gamma[p/\alpha]$ is the result of substituting α for all occurrences of p in γ.

Proof. Part (a) is proved by ADC (Section 2.6.). Consider for example $(\alpha \leftrightarrow \beta) \to (\gamma \to \alpha) \leftrightarrow (\gamma \to \beta)$. Analysis reduces the goal to two subgoals:

$$\alpha \leftrightarrow \beta, \gamma \to \alpha, \gamma \vdash \beta \quad \text{and} \quad \alpha \leftrightarrow \beta, \gamma \to \beta, \alpha, \gamma \vdash \alpha.$$

Direct chaining for the first subgoal is applied to $\alpha \leftrightarrow \beta, \gamma \to \alpha, \gamma$, and it yields $\alpha \to \beta, \beta \to \alpha, \gamma \to \alpha, \gamma, \alpha, \beta$ as required. The second subgoal is obtained from the first one by interchanging α and β.

Consider $\alpha \leftrightarrow \beta \rightarrow (\alpha \vee \gamma) \leftrightarrow (\beta \vee \gamma)$. Analysis yiels $\alpha \leftrightarrow \beta, \alpha \vee \gamma \vdash \beta \vee \gamma$ (and the result of interchanging β and α), and then:

$$\alpha \rightarrow \beta, \beta \rightarrow \alpha, \alpha \vdash \beta \vee \gamma \text{ and } \alpha \rightarrow \beta, \beta \rightarrow \alpha, \gamma \vdash \beta \vee \gamma.$$

The second subgoal is immediate by $\vee I$, and the first subgoal is obtained by direct chaining: $\alpha \rightarrow \beta, \alpha, \beta$.

(b) Use induction on γ with $\gamma \leftrightarrow \gamma$ in the induction base and Part (a) in the induction step. ⊣

2.9. Classical Propositional Logic

2.9.1. Semantics: Truth Tables

We recall here some standard elementary definitions. Admissible values for propositional variables in the standard semantics for CPC are *true* and *false*, often denoted by 1,0.

Truth values of compound formulas are computed from truth values of variables by the standard rules summarized by the following truth tables. The symbol $\top$ always takes the value 1, and $\bot$ always takes the value 0.

p	$\neg p$
1	0
0	1

p	q	$p \vee q$	$p \& q$	$p \rightarrow q$	$p \leftrightarrow q$
0	0	0	0	1	1
0	1	1	0	1	0
1	0	1	0	0	0
1	1	1	1	1	1

Every given assignment of truth values to variables occurring in a given formula (*truth value assignment*) determines the truth value of this formula.

EXAMPLE 2.16. Let $\alpha \equiv p \vee \neg p$. Then for every truth value assignment v we have $v(\alpha) = 1$: Consider cases $v(p) = 0$, $v(p) = 1$. So α is true, that is it takes value *true* under every truth value assignment. This means by definition that α is a *tautology* or a *valid formula* of CPC.

EXAMPLE 2.17. Let $\alpha \equiv \neg(p \vee q) \vee p$. Consider the assignment $v(p) = 0$ and $v(q) = 1$. Then:

$$v(\alpha) = v(\neg(p \vee q) \vee 0) = v(\neg(p \vee q)) = v(\neg(0 \vee 1)) = 0$$

Since α is false under a given assignment, it is not a tautology. The assignment $v(p) = 0, v(q) = 1$ is said to be a *falsifying assignment* for α. Assignment $v(p) = v(q) = 1$ gives $v(\alpha) = 1$, so it is a *verifying* (or *satisfying*) *assignment*.

Since operators $\vee, \&, \rightarrow, \neg$, and so on, defined in this way act on truth values of their arguments, they are called *truth functional operators* or *truth functional connectives*.

An operator with one argument (such as $\neg$) is called *monadic*; an operator with two arguments (such as $\vee$) is *binary*.

We can also state truth table definitions in abbreviated linear form. By checking against truth tables, it is easy to verify that:

$$v(\neg\alpha) = 1 - v(\alpha)$$

$$\begin{aligned} &v(\alpha \vee \beta) = \max(v(\alpha), v(\beta)) \\ &v(\alpha \vee 0) = v(\alpha) \\ &v(\alpha \vee 1) = 1 \end{aligned}$$

$$\begin{aligned} &v(\alpha \& \beta) = \min(v(\alpha), v(\beta)) \\ &v(\alpha \& 1) = v(\alpha) \\ &v(\alpha \& 0) = 0 \end{aligned}$$

$$\begin{aligned} &v(\alpha \to \beta) = \max(1 - v(\alpha), v(\beta)) \\ &v(1 \to \beta) = v(\beta) \\ &v(0 \to \beta) = 1 \end{aligned}$$

$$\begin{aligned} &v(\alpha \leftrightarrow \beta) = 1 \quad \text{iff} \quad v(\alpha) = v(\beta) \\ &v(\alpha \leftrightarrow 0) = v(\neg\alpha) \\ &v(\alpha \leftrightarrow 1) = v(\alpha) \end{aligned}$$

EXERCISE 2.2. Prove that $((p \leftrightarrow q) \leftrightarrow q) \leftrightarrow p$ is a tautology, but $(((p \leftrightarrow q) \leftrightarrow q) \leftrightarrow p) \leftrightarrow p$ is not.

Such tautologies as:

$$\neg p \leftrightarrow (p \to \bot), \qquad (p \leftrightarrow q) \leftrightarrow ((p \to q)\&(q \to p)$$

justify the treatment of $\neg, \leftrightarrow$ as defined connectives.

2.9.2. Logical Computations

Let us list some derivable formulas encoding truth tables. For a truth value $\sigma = 0, 1$ define

$$\alpha^\sigma \equiv \begin{cases} \alpha & \text{if } \sigma = 1 \\ \neg\alpha & \text{if } \sigma = 0 \end{cases}$$

LEMMA 2.11. *(a)* $\alpha^\sigma \vdash (\neg\alpha)^{\neg\sigma}$
(b) $\alpha^\sigma, \beta^\tau \vdash (\alpha \odot \beta)^{\odot(\sigma,\tau)}$ *for* $\odot = \vee, \&, \to, \leftrightarrow$

Proof.(a) Since $\neg 0 = 1$, $\neg 1 = 0$, we must prove that:

$$\alpha^1 \equiv \alpha \vdash (\neg\alpha)^0 \equiv \neg\neg\alpha \text{ and } \alpha^0 \equiv \neg\alpha \vdash (\neg\alpha)^1 \equiv \neg\alpha.$$

The second sequent is an axiom, and the first is obtained by ADC.

(b) Let us list (slightly strengthened) goals in more detail:

$$\alpha,\beta\vdash\alpha\&\beta \qquad \neg\alpha\vdash\neg(\alpha\&\beta) \qquad \neg\beta\vdash\neg(\alpha\&\beta)$$

$$\alpha\vdash\alpha\vee\beta \qquad \beta\vdash\alpha\vee\beta \qquad \neg\alpha,\neg\beta\vdash\neg(\alpha\vee\beta)$$

$$\beta\vdash\alpha\to\beta \qquad \neg\alpha\vdash\alpha\to\beta \qquad \alpha,\neg\beta\vdash\alpha\to\beta$$

$$\alpha,\beta\vdash\alpha\leftrightarrow\beta \qquad \neg\alpha,\neg\beta\vdash\alpha\leftrightarrow\beta \qquad \alpha,\neg\beta\vdash\neg(\alpha\leftrightarrow\beta) \qquad \neg\alpha,\beta\vdash\neg(\alpha\leftrightarrow\beta)$$

Now use ADC. ⊣

THEOREM 2.1. *Assume that all propositional variables of a formula ϕ are among $p_1,\ldots,p_n$, and let v be a truth value assignment to $p_1,\ldots,p_n$. Then:*

$$p_1^{v(p_1)},\ldots,p_n^{v(p_n)}\vdash\phi^{v(\phi)} \qquad or \qquad \mathbf{p}^{v(\mathbf{p})}\vdash\phi^{v(\phi)} \tag{2.4}$$

Proof. We use induction on ϕ. If $\phi=p_i$, then (2.4) is an axiom. If $\phi\equiv\phi_1\odot\phi_2$, then by IH:

$$\mathbf{p}^{v(\mathbf{p})}\vdash\phi_1^{v(\phi_1)}, \quad \mathbf{p}^{v(\mathbf{p})}\vdash\phi_2^{v(\phi_2)}.$$

By Lemma 2.9.2$\phi_1^{v(\phi_1)},\phi_2^{v(\phi_2)}\vdash(\phi_1\odot\phi_2)^{\odot(v(\phi_1),v(\phi_2))}\equiv\phi^{v(\phi)}$, which proves the induction step. ⊣

THEOREM 2.2. *(a) Every tautology is derivable in NKp;*
(b) $\vdash\neg\neg\tau$ in NJp for every tautology τ.

Proof. Consider Part (b) first. Let τ be a tautology, that is, τ is true under any truth value assignment v : $v(\tau)=1$. Then (2.4) takes the form $p_1^{v(p_1)},\ldots,p_n^{v(p_n)}\vdash\tau$, which with $\tau\to\neg\neg\tau$ [Lemma 2.7.(b1)] implies that:

$$p_1^{v(p_1)},\ldots,p_n^{v(p_n)}\vdash\neg\neg\tau. \tag{2.5}$$

Let us prove in NJ that for any k, $0\le k\le n$

$$p_1^{v(p_1)},\ldots,p_k^{v(p_k)}\vdash\neg\neg\tau \quad \text{for every } v \tag{2.6}$$

by induction on $i=n-k$. Then $k=0$ yields $\vdash\neg\neg\tau$. Induction base $i=0$, that is, $k=n$ is (2.5). The induction step, that is, the passage from (2.6) to:

$$p_1^{v(p_1)},\ldots,p_{k-1}^{v(p_{k-1})}\vdash\neg\neg\tau \quad \text{for every } v \tag{2.7}$$

is proved as follows. Take any v (assumed to be defined only at $p_1,\ldots,p_{k-1}$) and consider $v^1\equiv v\cup\{p_k=1\}$, $v^0\equiv v\cup\{p_k=0\}$. By IH:

$$p_1^{v(p_1)},\ldots,p_{k-1}^{v(p_{k-1})},p_k\vdash\neg\neg\tau \quad \text{and} \quad p_1^{v(p_1)},\ldots,p_{k-1}^{v(p_{k-1})},\neg p_k\vdash\neg\neg\tau$$

which implies

$$p_1^{v(p_1)},\ldots,p_{k-1}^{v(p_{k-1})}\vdash\neg\neg\neg\neg\tau$$

by (2.2) and hence (2.7) by Lemma 2.7.(b2), as required.

To get (a) from (b), note that $\neg\neg\alpha\Rightarrow\alpha$ is derivable in NKp by the $\bot_c$-rule. ⊣

Chapter 3

Negative Translation: Glivenko's Theorem

THEOREM 3.1. *(Glivenko's theorem)* $\Gamma \vdash \neg\alpha$ *iff* $\&\Gamma \rightarrow \neg\alpha$ *is a tautology. In particular a formula beginning with a negation is derivable in NJp iff it is a tautology.*

Proof. By Lemma 2.7., every derivable sequent is a tautology. In the opposite direction, if a formula $\&\Gamma \rightarrow \neg\alpha$ is a tautology, then by Theorem 2.9.2(b), $\vdash \neg\neg(\&\Gamma \rightarrow \neg\alpha)$. Now use the implication:

$$\neg\neg(\gamma \rightarrow \neg\alpha) \rightarrow (\gamma \rightarrow \neg\alpha),$$

which is obtained from an instance of (2.4): $\gamma, \alpha \Rightarrow \neg(\gamma \rightarrow \neg\alpha)$. ⊣

The remaining part of this Chapter shows that it is possible to embed classical logic NKp into intuitionistic system NJp by inserting double negation to turn off constructive content of disjunctions and atomic formulas (which stand for arbitrary sentences and may potentially have constructive content).

DEFINITION 3.1. *Define inductively operation* ^{neg} *transforming formulas into formulas:*

$$p^{neg} := \neg\neg p \quad \textit{for atomic } p$$

$$(\alpha\&\beta)^{neg} := \alpha^{neg}\&\beta^{neg} \qquad (\alpha \rightarrow \beta)^{neg} := \alpha^{neg} \rightarrow \beta^{neg}$$

$$(\alpha \vee \beta)^{neg} := \neg(\neg\alpha^{neg}\&\neg\beta^{neg})$$

In connection with the last clause, note that de Morgan's law [Lemma 2.7.(c)] implies that:

$$\vdash \neg(\neg\alpha\&\neg\beta) \leftrightarrow \neg\neg(\alpha \vee \beta).$$

Note also that:

$$\vdash (\neg\alpha)^{neg} \leftrightarrow \neg(\alpha^{neg}),$$

since $(\neg\alpha)^{neg} \equiv (\alpha^{neg} \rightarrow \neg\neg\bot) \leftrightarrow (\alpha^{neg} \rightarrow \bot)$.

The next Lemma is used to justify the negative translation.

DEFINITION 3.2. *A propositional formula is* negative *if it does not contain* $\vee$, *and all atomic subformula are negated.*

LEMMA 3.1.

$$(a)\qquad \vdash \neg\neg(\alpha\&\beta) \leftrightarrow (\neg\neg\alpha\&\neg\neg\beta) \qquad \vdash \neg\neg(\alpha \to \beta) \leftrightarrow (\neg\neg\alpha \to \neg\neg\beta)$$

(b) $\vdash \neg\neg\alpha \leftrightarrow \alpha$ *for every negative formula* α.

Proof. Part (a): Use ADC and Glivenko's theorem.

Part (b): We use induction on α, Lemma 2.7.(b1),(b2), and (a). For the induction base, we begin with negations of atomic formulas: $\neg\neg(\neg p) \leftrightarrow \neg p$.

Induction step. For $\odot \equiv \&, \to$, $\alpha \equiv \phi \odot \psi$, one has

$$\begin{aligned}\neg\neg\alpha \equiv\ & \neg\neg(\phi \odot \psi) \leftrightarrow (\neg\neg\phi \odot \neg\neg\psi)\\ & \leftrightarrow (\phi \odot \psi) \text{ (by IH).}\end{aligned}$$

This concluses the proof since $\vee$ does not occur in α. ⊣

LEMMA 3.2. *Every rule*

$$\frac{\Gamma \Rightarrow \phi \quad \Delta \Rightarrow \psi}{\Sigma \Rightarrow \theta}$$

of NKp is stable under Gödel's negative translation. That is, the rule:

$$\frac{\Gamma^{neg} \Rightarrow \phi^{neg} \quad \Delta^{neg} \Rightarrow \psi^{neg}}{\Sigma^{neg} \Rightarrow \theta^{neg}}$$

is derivable and similarly for one-premise and three-premise rules.

Proof. Translations of the rules $\&I, \&E, \to I, \to E, \bot_i$ are instances of the same rules. Translation of $\bot_c$

$$\frac{\neg\alpha^{neg}, \Gamma^{neg} \Rightarrow \bot}{\Gamma^{neg} \Rightarrow \alpha^{neg}}$$

is handled by Lemma 3(b). For the rules for $\vee$, it is enough to establish the following:

$$\alpha_i \vdash \neg(\neg\alpha_0\&\neg\alpha_1) \qquad (i = 0, 1) \tag{3.1}$$

$$\neg(\neg\phi\&\neg\psi), \phi \to \theta^{neg}, \psi \to \theta^{neg} \vdash \theta^{neg} \tag{3.2}$$

Relation (3.1) is obtained by ADC. To obtain (3.2), prefix the last θ^{neg} by $\neg\neg$ (Lemma 3) and reduce this goal to a subgoal:

$$\phi \to \theta^{neg}, \psi \to \theta^{neg}, \neg\theta^{neg} \vdash (\neg\phi\&\neg\psi),$$

which is proved by ADC. ⊣

THEOREM 3.2. *A sequent* $\Gamma \Rightarrow \Delta$ *is derivable in NKp iff* $\Gamma^{neg} \Rightarrow \Delta^{neg}$ *is derivable in NJp.*

Proof. Easy direction: If $\Gamma^{neg} \Rightarrow \Delta^{neg}$ is derivable in NJp, it is derivable in NKp. Removing all double negations, we obtain $\Gamma \Rightarrow \Delta$ in NKp.

For the opposite direction, use Lemma 3. ⊣

Chapter 4

Program Interpretation of Intuitionistic Logic

4.1. BHK-Interpretation

Let us recall Brouwer's requirement of effective existence: To claim $\exists x\alpha$, we must point out an object t and justify $\alpha[x/t]$. To claim a disjunction $\alpha_0 \vee \alpha_1$, we must point out which of the disjuncts is true, that is, select an $i \in \{0, 1\}$ and justify α_i.

We describe now a "programming language" and a method of mechanically extracting programs in this language from intuitionistic proofs. This language is a version of simply typed lambda calculus [2]. It is based on a semantics of intuitionistic logic in terms of constructions, presupposing that any formula A encodes a problem, and corresponding construction c solves this problem or *realizes* A. We give an informal explanation of semantics of logical connectives that evolved from the work of Brouwer, Heyting, and Kolmogorov. It is called *BHK-interpretation.* We write $c\mathbf{r}\alpha$ for "c realizes α" or "c is a construction for α":

$c\mathbf{r}(\alpha_0 \wedge \alpha_1)$ iff c is a pair $c = \mathbf{p}(a_0, a_1)$ and $a_0\mathbf{r}\alpha_0$ and $a_1\mathbf{r}a_1$,
$c\mathbf{r}(\alpha_0 \vee \alpha_1)$ iff ($c = \mathbf{k}_0 a$ and $a\mathbf{r}\alpha_0$) or ($c = \mathbf{k}_1 a$ and $a\mathbf{r}\alpha_1$),
$c\mathbf{r}(\alpha \to \beta)$ iff c is a function and for every d, if $d\mathbf{r}\alpha$ then $c(d)\mathbf{r}\beta$,
not $c\mathbf{r}\bot$.

The last clause is equivalent to:

$c\mathbf{r}\neg\alpha$ iff $\neg d\mathbf{r}\alpha$ for every d.

We see that our language for such constructions should include pairing, projections (components of pairs), injection into a direct sum, forming functions, and applying functions to arguments.

EXAMPLE 4.1. Realization t of $(\alpha_0 \vee \alpha_1) \to (\beta_0 \vee \beta_1)$ is a program that for every pair $x = \mathbf{p}(i, a)$ such that a realizes α_i, produces a pair $t(x) = \mathbf{p}(j, b)$ such that b realizes β_j.

EXERCISE 4.1. What are realizations of formulas of the following form:

$$\alpha \to (\beta \& \gamma); \qquad (\alpha_0 \vee \alpha_1) \to \beta \& (\beta_0 \vee \beta_1); \qquad \alpha \to (\alpha_0 \vee \alpha_1) \to (\beta_0 \vee \beta_1)$$

4.2. Assignment $\mathcal{T}$ of Deductive Terms

This Section presents a language for writing realizations of formulas derivable in NJp. Basic constructs of this language are *pairing* $\mathbf{p}$ with projections $\mathbf{p}_0, \mathbf{p}_1$ satisfying:

$$\mathbf{p}_i(\mathbf{p}(t_0, t_1)) = t_i, \qquad i = 0, 1 \tag{4.1}$$

and lambda abstraction providing explicit definitions:

$$(\lambda x.t)(u) = t[x/u], \tag{4.2}$$

where $t[x/u]$ stands for the result of substituting a term u for all free occurrences of a variable x (of the same type) in t. In other words, if t is an expression containing a variable x, it is possible to define a function (denoted by $\lambda x.t$) that outputs the value $t[x/u]$ when the value of x is set to u. Sometimes we omit a dot and write λxt.

Application of a function t to an argument u is denoted by $t(u)$. Two more constructions are connected with disjunction and case distinction.

To indicate that an object t is to be treated as a realization of ϕ_0 in $\phi_0 \vee \phi_1$, we write $\mathbf{k}_0 t$, and similarly for $\mathbf{k}_1 t$.

Three-place operation D expresses case distinction for disjunction: If $u = \mathbf{k}_i t$ ($i \in \{0, 1\}$) is a realization of a disjunction $\phi_0 \vee \phi_1$ and t_0, t_1 are realizations for some formula θ depending on realizations x_0, x_1 of ϕ_0, ϕ_1, respectively, then $D_{x_0,x_1}(u, t_0, t_1)$ is a realization of ϕ_i formalizing the instruction: If t in $u = \mathbf{k}_i t$ realizes ϕ_0 (that is, $i = 0$), then the result is $t_0[x/t]$; otherwise $t_1[x/t]$. This can be expressed by an equality:

$$D_{x_0,x_1}(\mathbf{k}_i t, t_0, t_1) = t_i[x_i/t], \; i = 0, 1. \tag{4.3}$$

Operation $\perp_\phi$ provides a trivial realization of a formula ϕ under the assumption that a contradiction was obtained.

To formalize the assumption that realizations of some of the formulas are given, we assume for every formula ϕ of the language under consideration a countably infinite supply of variables of type ϕ. For distinct formulas ϕ, the corresponding sets of variables are disjoint, and the set of typed variables is disjoint from the set of individual variables. We use x^ϕ, y^ϕ, z^ϕ for arbitrary variables of type ϕ; we omit the type superscript when clear from the context.

Now we assign a term $\mathcal{T}(d)$ to every natural deduction d deriving a sequent:

$$\Gamma \Rightarrow \phi. \tag{4.4}$$

A term u assigned to a derivable sequent (4.4) is supposed to realize a formula ϕ according to the BHK-interpretation under assumptions Γ. We sometimes write t^ϕ to stress this. To reflect dependence of assumptions, every assignment depends on a *context* that is itself an assignment:

$$z^{\phi_1} : \phi_1, \ldots, z^{\phi_n} : \phi_n$$

of distinct typed variables to formulas in $\Gamma \equiv \phi_1, \ldots, \phi_n$ written sometimes as $\mathbf{z} : \Gamma$. These typed variables stand for hypothetical realizations of the assumptions Γ. When a term u is assigned to ϕ, the sequent (4.4) is transformed into a statement:

$$z^{\phi_1} : \phi_1, \ldots, z^{\phi_n} : \phi_n \Rightarrow \; u : \phi \qquad \text{or} \qquad \mathbf{z} : \Gamma \Rightarrow u : \phi$$

Contexts are treated as sets, not multisets. In particular $\mathbf{z} : \Gamma, \mathbf{z}' : \Delta$ stands for the union of sets, and $\mathbf{z} : \Gamma, \mathbf{z} : \Gamma \equiv \mathbf{z} : \Gamma$.

Deductive terms and the assignment of a term to a deduction is defined inductively. Assignments for axioms are given explicitly, and for every logical inference rule, there is an operation that transforms assignments for the premises into an assignment for the conclusion of the rule.

4.2.1. Assignment Rules

Axioms:

$$x : \phi \Rightarrow x : \phi$$

Inference rules:

$$\frac{\mathbf{z} : \Gamma \Rightarrow t : \phi \quad \mathbf{z}' : \Delta \Rightarrow u : \psi}{\mathbf{z} : \Gamma, \mathbf{z}' : \Delta \Rightarrow \mathbf{p}(t, u) : (\phi \& \psi)} \,\&I \qquad \frac{\mathbf{z} : \Gamma \Rightarrow t : \phi_0 \& \phi_1}{\mathbf{z} : \Gamma \Rightarrow \mathbf{p}_i t : \phi_i} \,\&E \qquad i = 0, 1$$

$$\frac{\mathbf{z} : \Gamma \Rightarrow t : \phi_i}{\mathbf{z} : \Gamma, \Rightarrow \mathbf{k}_i t : (\phi_0 \vee \phi_1)} \,\vee I \qquad i = 0, 1$$

$$\frac{\mathbf{z} : \Gamma \Rightarrow t : \phi \vee \psi \quad (x^\phi : \phi)^0, \mathbf{z}' : \Delta \Rightarrow t_0 : \theta \quad (y^\psi : \psi)^0, \mathbf{z}'' : \Sigma \Rightarrow t_1 : \theta}{\mathbf{z} : \Gamma, \mathbf{z}' : \Delta, \mathbf{z}'' : \Sigma \Rightarrow D_{xy}(t, t_0, t_1) : \theta} \,\vee E$$

$$\frac{\mathbf{z} : \Gamma \Rightarrow t : (\phi \to \psi) \quad \mathbf{z}' : \Delta \Rightarrow u : \phi}{\mathbf{z} : \Gamma, \mathbf{z}' : \Delta \Rightarrow t(u) : \psi} \to E \qquad \frac{(x : \phi)^0, \mathbf{z} : \Gamma \Rightarrow t : \psi}{\mathbf{z} : \Gamma \Rightarrow \lambda x.t : (\phi \to \psi)} \to I$$

$$\frac{\mathbf{z} : \Gamma \Rightarrow t : \bot}{\mathbf{z} : \Gamma \Rightarrow \bot_\phi(t) : \phi} \,\bot_i$$

Term assignment $\mathcal{T}(d)$ to a natural deduction d is defined in a standard way by application of the term assignment rules. Notation $\Gamma \Rightarrow \; t : \alpha$ or $\mathbf{z} : \Gamma \Rightarrow \; t : \alpha$ means that $t = \mathcal{T}(d)$ for some natural deduction $d : \Gamma \Rightarrow \alpha$.

The symbol λx binds variable x, and D_{xy} binds variables x and y. *Free variables* of a term are defined in a familiar way:

$$FV(x) := x; \quad FV(\mathbf{p}_i t) := FV(\mathbf{k}_i t) := FV(\bot_\phi(t)) := FV(t)$$

$$FV(\mathbf{p}(t,u)) := FV(t(u)) := FV(t) \cup FV(u); \quad FV(\lambda xt) := FV(t) - \{x\}$$
$$FV(D_{xy}(t,t_0,t_1)) :== FV(t) \cup FV(t_0) \cup FV(t_1) - \{x,y\}.$$

The term assignment can be extended to structural rules so that weakening does not change term assignment, and contraction identifies variables for contracted assumptions:

$$\frac{\mathbf{z}:\Gamma \Rightarrow u:\phi}{x:\psi,\mathbf{z}:\Gamma \Rightarrow u:\phi}\; weak \qquad \frac{x:\psi, y:\psi, \mathbf{z}:\Gamma \Rightarrow u:\phi}{x:\psi,\mathbf{z}:\Gamma \Rightarrow u[y/x]:\phi}\; contr$$

One simple and sufficiently general way of finding a realization of a formula is to derive it in NJp and compute the assigned term.

EXAMPLE 4.2. Find a term t realizing $(p \to q) \to ((q \to r) \to (p \to r))$. Consider the following natural deduction d:

$$\dfrac{\dfrac{\dfrac{\dfrac{q \to r \Rightarrow q \to r \qquad \dfrac{p \to q \Rightarrow p \to q \quad p \Rightarrow p}{p, p \to q \Rightarrow q}}{p \to q, q \to r, p \Rightarrow r}}{p \to q, q \to r \Rightarrow p \to r}}{p \to q \Rightarrow (q \to r) \to (p \to r)}}{\Rightarrow (p \to q) \to ((q \to r) \to (p \to r))}$$

Assign terms:

$$\dfrac{\dfrac{\dfrac{\dfrac{y:q \to r \Rightarrow y:q \to r \qquad \dfrac{x:p \to q \Rightarrow x:p \to q \quad z:p \Rightarrow z:p}{z:p, x:p \to q \Rightarrow x(z):q}}{x:p \to q, y:q \to r, z:p \Rightarrow y(x(z)):r}}{x:p \to q, y:q \to r \Rightarrow \lambda z^p.y(x(z)):p \to r}}{x:p \to q \Rightarrow \lambda y^{q \to r}\lambda z^p.y(x(z)):(q \to r) \to (p \to r)}}{\Rightarrow \lambda x^{p \to q}\lambda y^{q \to r}\lambda z^p.y(x(z)):(p \to q) \to ((q \to r) \to (p \to r))}$$

Hence $\mathcal{T}(d) \equiv \lambda x \lambda y \lambda z.y(x(z))$.

Several more examples are

$$\Rightarrow \lambda x^p.x : (p \to p)$$

$$\Rightarrow \lambda x^p \lambda y^q.x : p \to (q \to p)$$

$$\Rightarrow \lambda x^{p_i}.\mathbf{k}_i x : (p_i \to p_0 \vee p_1)$$

$$x:p \to r, y:q \to r, z:p \vee q \Rightarrow D_{u^p,v^q}(z, x(u^p), y(v^q)):r$$

Hence:

$$\Rightarrow \lambda x \lambda y \lambda z.D_{u^p,v^q}(z, x(u^p), y(v^q)) : (p \to q) \to ((q \to r) \to (p \vee q \to r))$$

EXERCISE 4.2. Confirm the preceding realizations and find realizations for the following formulas using deductions in NJp:

$$p\&q \rightarrow p, \; p\&q \rightarrow q, \; p \rightarrow (q \rightarrow p\&q)$$

$$(p \rightarrow q) \rightarrow ((p \rightarrow (q \rightarrow r)) \rightarrow (p \rightarrow r))$$

4.3. Properties of Term Assignment $\mathcal{T}$

The $\mathcal{T}(d)$ is defined up to renaming of free variables assigned to axioms. If $d : \Gamma \Rightarrow \phi$, that is, d is a deduction of $\Gamma \Rightarrow \phi$, then $\mathbf{z} : \Gamma \Rightarrow \mathcal{T}(d) : \phi$.

In fact the operation $\mathcal{T}$ is an isomorphism: It has an inverse operation (denoted by $\mathcal{D}$ below) preserving both syntactic identity and more important relation of β-equality (see Chapter 5 below).

DEFINITION 4.1. *(operation $\mathcal{D}$). For every deductive term t^ϕ with $FV(t) = \mathbf{z}^\Gamma$ we define a deduction:*

$$\mathcal{D}(t) : \; \Gamma \Rightarrow \phi \tag{4.5}$$

If $t^\phi \equiv x^\phi$ then $\mathcal{D}(t) := \; \phi \Rightarrow \phi$ (Axiom)
If $t^{\phi_i} \equiv \mathbf{p}_i u^{\phi_0 \& \phi_1}$ then $\mathcal{D}(t)$ is obtained from $\mathcal{D}(u)$ by $\&E$:

$$\frac{\mathcal{D}(u) : \; \Gamma \Rightarrow \phi_0 \& \phi_1}{\mathcal{D}(\mathbf{p}_i u) : \; \Gamma \Rightarrow \phi_i} \; \&E$$

If $t^{\phi\&\psi} \equiv \mathbf{p}(u^\phi, v^\psi)$ with $FV(u) \equiv \mathbf{z}^\Sigma$, $FV(v) \equiv \mathbf{z}'^\Delta$, then:

$$\frac{\mathcal{D}(u) : \; \Sigma \Rightarrow \phi \quad \mathcal{D}(v) : \; \Delta \Rightarrow \psi}{\mathcal{D}(\mathbf{p}(u, v)) : \; [\Sigma, \Delta] \Rightarrow \phi\&\psi} \; \&I,$$

where occurrences of identical assumptions in Σ and Δ are identified in $[\Sigma, \Delta]$ exactly when these occurrences are assigned the same variable in the contexts $\mathbf{z} : \Sigma$ and $\mathbf{z}' : \Delta$.
If $t^\psi \equiv u^{\phi\rightarrow\psi}(v^\phi)$, then $\mathcal{D}(t)$ is obtained from $\mathcal{D}(u), \mathcal{D}(v)$ by $\rightarrow E$ with the same identification of assumptions as in the previous case.
If $t^{\phi\rightarrow\psi} \equiv \lambda x^\phi . u^\psi$, then:

$$\frac{\mathcal{D}(u) : \; (\phi)^0, \Gamma \Rightarrow \psi}{\mathcal{D}(\lambda x^\phi . u) : \; \Gamma \Rightarrow \phi \rightarrow \psi} \rightarrow I,$$

where assumption ϕ is present in the premise iff $x^\phi \in FV(u)$.
If $t^{\phi_0 \vee \phi_1} \equiv \mathbf{k}_i u^{\phi_i}$, then:

$$\frac{\mathcal{D}(u) : \; \Gamma \Rightarrow \phi_i}{\mathcal{D}(\mathbf{k}_i u) : \; \Gamma \Rightarrow \phi_0 \vee \phi_1} \vee I.$$

If $t^\theta \equiv D_{x^\phi, y^\psi}(u^{\phi\vee\psi}, t_0^\theta, t_1^\theta)$, *then:*

$$\frac{\mathcal{D}(u) : \ \Gamma \Rightarrow \phi \vee \psi \quad \mathcal{D}(t_0) : \ (\phi)^0, \Delta \Rightarrow \theta \quad \mathcal{D}(t_1) : \ (\psi)^0, \Sigma \Rightarrow \theta}{\mathcal{D}(D_{x,y}(u, t_0, t_1)) : \ [\Gamma, \Delta, \Sigma] \Rightarrow \theta} \vee E,$$

with the same identifications of formulas in $[\Gamma, \Delta, \Sigma]$ *as before. Assumption* ϕ *is present in the second premise iff* x^ϕ *is free in* t_0*, and assumption* ψ *is present in the third premise iff* y^ψ *is free in* t_1*.*

LEMMA 4.1. *Up to renaming of free and bound variables,*

(a) $\mathcal{D}(\mathcal{T}(d)) \equiv d$ *for every deduction* d

(b) $\mathcal{T}(\mathcal{D}(t)) \equiv t$ *for every deductive term* t

Proof. We establish Parts (a),(b) by a simultaneous induction on d, t. The induction base (axiom d or variable $t \equiv x^\phi$) is obvious. The induction step is proved by cases corresponding to each of the inference rules. Consider one of them:

(a) $d : \Gamma \Rightarrow \phi_i$ is obtained from $e : \Gamma \Rightarrow \phi_0 \& \phi_1$ by $\& E$. Then $\mathcal{T}(d) \equiv \mathbf{p}_i \mathcal{T}(e)$ by the definition of $\mathcal{T}$, and:

$$\frac{\mathcal{D}(\mathcal{T}(e)) : \ \Gamma \Rightarrow \phi_0 \& \phi_1}{\mathcal{D}(\mathcal{T}(d)) : \ \Gamma \Rightarrow \phi_i} \& E$$

by the definition of $\mathcal{D}$. But the latter figure coincides with d, since $\mathcal{D}(\mathcal{T}(e)) \equiv e$ by IH.

(b) $t^{\phi_i} \equiv \mathbf{p}_i s^{\phi_0 \& \phi_1}$. Then $\mathcal{D}(t)$ is obtained from $\mathcal{D}(s)$ by $\& E$ and $\mathcal{T}(\mathcal{D}(t)) \equiv \mathbf{p}_i \mathcal{T}(\mathcal{D}(s)) \equiv \mathbf{p}_i s$ (by IH) $\equiv t$ as required.

Other cases are similar. ⊣

Chapter 5

Computations with Deductions

Let us recall that the conclusion of the rule $\bot_i$ is assumed to be atomic.

5.1. Conversions and Reductions of Deductive Terms

Relations (4.1–4.3) are naturally treated as computation rules that simplify the left-hand side into the right-hand side. In other words, an operational semantics for the language of terms is given by the following term *conversion* (rewriting) rules:

$$(\lambda x.t)(t') \quad conv \quad t[x/t'] \tag{5.1}$$

$$\mathbf{p}_i(\mathbf{p}(t_0, t_1)) \quad conv \quad t_i \quad i = 0, 1 \tag{5.2}$$

$$D_{x_0,x_1}(\mathbf{k}_i t, t_0, t_1) \quad conv \quad t_i[x_i/t],\ i = 0, 1 \tag{5.3}$$

These relations are called *β-conversions.* [Originally the term referred only to (5.1)].

One-step reduction red_1 is a conversion of a subterm:

$$\text{if } u \ conv \ u' \ \text{ then } \ t[x/u] \ red_1 \ t[x/u']. \tag{5.4}$$

Here u is a *redex* and u' is a *reductum.*

The relation *red* is a transitive reflexive closure of red_1: $t \ red \ t'$ if there is a *reduction sequence*:

$$t \equiv t_0, \ldots, t_n \equiv t' \quad (n \geq 0)$$

such that $t_i \ red_1 \ t_{i+1}$ for every $i < n$.

A term t is *in normal form* or t *is normal* if it does not contain a redex; t *has a normal form* if there is a normal s such that $t \ red \ s$. Reduction sequence is an analog of a computation, and a normal form is an analog of a value.

5.2. Conversions and Reductions of Natural Deductions

Let us describe transformations of natural deductions corresponding to a reduction of terms. Each of these transformations converts an occurrence of an introduction rule immediately followed by an elimination of the introduced connective. Such a pair of inferences is called a *cut* in this Section. We do not elaborate here on a connection with the *cut rule* of Example 2.5..

The &-conversion corresponding to the pairing conversion (5.2):

$$\cfrac{\cfrac{d_0 : \Gamma \Rightarrow \phi_0 \quad d_1 : \Delta \Rightarrow \phi_1}{[\Gamma, \Delta] \Rightarrow \phi_0 \& \phi_1}\,\&I}{d : [\Gamma, \Delta] \to \phi_i}\,\&E \qquad conv \qquad d_0 : \Gamma \Rightarrow \phi_0$$

Note that conversion can change the set of assumptions.

A description of remaining conversions uses a ***substitution operation for natural deductions.*** Let us recall that in every inference (application of an inference rule), each occurrence of an undischarged assumption formula in a premise is represented by an occurrence of the same formula in the conclusion, which is called its ***immediate descendant*** in the conclusion. For any natural deduction, starting with an assumption formula in the antecedent of one of the sequents, we are led successively to unique ***descendants*** of this occurrence in the sequents below it. The chain of such descendants stops at discharged assumptions. ***Ancestors*** of a given formula (occurrence) are occurrences that have it as a descendant. We say that a given antecedent formula (occurrence) is ***traceable*** to any of its ancestors (including itself). Each occurrence has at most one descendant in a given sequent. It is important to note that all ancestors of a given (occurrence of) assumption are assigned one and the same variable in the assignment of deductive terms to deductions.

EXAMPLE 5.1. Consider the following deduction of a sequent $\delta \to q \Rightarrow q$, where $\delta \equiv p \vee (p \to q)$.

$$\dfrac{\underline{\delta \to q} \Rightarrow \delta \to q \qquad \dfrac{\dfrac{\dfrac{\underline{\delta \to q} \Rightarrow \delta \to q \quad \dfrac{p \Rightarrow p}{p \Rightarrow \delta}}{\underline{\delta \to q}, p \Rightarrow q}}{\underline{\delta \to q} \Rightarrow p \to q}}{\underline{\delta \to q} \Rightarrow \delta}}{\underline{\delta \to q} \Rightarrow q}$$

Underlined occurrences of the assumption $\delta \to q$ are ancestors of the lowermost occurrence of this formula.

For given deductions d and d' of sequents $\alpha, \Gamma \Rightarrow \beta$, and $\Delta \Rightarrow \alpha$, the ***result of substituting d' for α into d*** is obtained by replacing all ancestors of α in d by Δ and writing d' over former axioms $\alpha \Rightarrow \alpha$. Taking deductive terms into consideration and writing the result of substitution at the right, yield the following:

$$\begin{array}{ccc} & x:\alpha\Rightarrow x:\alpha & \mathbf{y}:\Delta\Rightarrow s:\alpha \\ & \nwarrow\uparrow\nearrow & \nwarrow\uparrow\nearrow \\ & x:\alpha,\mathbf{z}':\Gamma'\Rightarrow t':\beta' & \mathbf{z}',\mathbf{y}:[\Gamma',\Delta]\Rightarrow t'[x/s]:\beta' \\ & \nwarrow\uparrow\nearrow & \nwarrow\uparrow\nearrow \\ d':\ \mathbf{y}:\Delta\Rightarrow s:\alpha & d:\ x:\alpha,\mathbf{z}:\Gamma\Rightarrow t:\beta & \mathbf{z},\mathbf{y}:[\Gamma,\Delta]\Rightarrow t[x/s]:\beta \end{array}$$

The arrows $\nwarrow\uparrow\nearrow$ show possible branching of the deduction at the binary and ternary rules ($\&I, \rightarrow E, \vee E$).

LEMMA 5.1.

(a) All inference rules are preserved by substitution.

(b) Operations $\mathcal{T}, \mathcal{D}$ commute with substitution under suitable proviso to avoid collision of bound variables.

($b\mathcal{T}$) If a deduction e is the result of substituting a deduction $d':\ \Delta\Rightarrow\alpha$ for the assumption (occurrence) $x:\alpha$ into a deduction $d:\ x:\alpha,\Gamma\Rightarrow\beta$, then:

$$\mathcal{T}(e)\equiv\mathcal{T}(d)[x/\mathcal{T}(d')] \tag{5.5}$$

($b\mathcal{D}$) The deduction $\mathcal{D}(t^\beta[x^\alpha/s^\alpha])$ is the result of substituting a deduction $\mathcal{D}(s)$: $\Delta\Rightarrow\alpha$ for the assumption (occurrence) $x:\alpha$ into a deduction

$\mathcal{D}(t):\ x:\alpha,\Gamma\Rightarrow\beta$.

Proof. Check the statement for each rule of NJp and apply induction on the length of deduction. $\dashv$

The $\rightarrow$-*conversion* is now defined as follows:

$$\begin{array}{c} u:\alpha\Rightarrow u:\alpha \\ \nwarrow\uparrow\nearrow \\ \mathbf{z}':\Gamma',u:\alpha\Rightarrow t':\beta' \\ \nwarrow\uparrow\nearrow \\ \dfrac{\dfrac{\mathbf{z}:\Gamma,u:\alpha\Rightarrow t:\beta}{\mathbf{z}:\Gamma\Rightarrow\lambda u.t:\alpha\rightarrow\beta}\quad \mathbf{y}:\Delta\Rightarrow s:\alpha}{\mathbf{z},\mathbf{y}:[\Gamma,\Delta]\Rightarrow(\lambda u.t)s:\beta} \end{array} \quad conv \quad \begin{array}{c} \mathbf{y}:\Delta\Rightarrow u:\alpha \\ \nwarrow\uparrow\nearrow \\ \mathbf{z}',\mathbf{y}:[\Gamma',\Delta]\Rightarrow t'[u/s]:\beta' \\ \nwarrow\uparrow\nearrow \\ \mathbf{z},\mathbf{y}:[\Gamma,\Delta]\Rightarrow t[u/s]:\beta \end{array}$$

The result of conversion is obtained from the derivation of the premise of $\rightarrow I$ in the original derivation by substitution. If there is no dependence on the assumption α in the $\rightarrow$-introduction, then the result of conversion is just the given derivation of $\Gamma\Rightarrow\beta$.

The $\vee$-conversion:

$$\begin{array}{c} x^\phi:\phi\Rightarrow x^\phi:\phi \\ \nwarrow\uparrow\nearrow \\ x^\phi:\phi,\mathbf{z}_1:\Delta'\Rightarrow t_0':\theta' \\ \nwarrow\uparrow\nearrow \end{array}$$

$$\dfrac{\dfrac{\mathbf{z}:\Gamma\Rightarrow t:\phi}{\mathbf{z}:\Gamma\Rightarrow\mathbf{k}_0 t:\phi\vee\psi}\quad x^\phi:\phi,\mathbf{z}':\Delta\Rightarrow t_0:\theta\quad \mathbf{y}^\psi:\psi,\mathbf{z}'':\Sigma\Rightarrow t_1:\theta}{\mathbf{z},\mathbf{z}',\mathbf{z}'':[\Gamma,\Delta,\Sigma]\Rightarrow D_{x,y}(\mathbf{k}_0 t,t_0,t_1):\theta}$$

$$\begin{array}{ll} & \mathbf{z} : \Gamma \Rightarrow t^{\phi} : \phi \\ & \nwarrow\uparrow\nearrow \\ & \mathbf{z} : \Gamma, \mathbf{z}_1 : \Delta' \Rightarrow t_0'[x/t] : \theta' \\ & \nwarrow\uparrow\nearrow \\ conv & [\mathbf{z}, \mathbf{z}'] : [\Gamma, \Delta] \Rightarrow t_0[x/t] : \theta \end{array}$$

The result of conversion is obtained from the derivation of the second premise of the $\vee E$ in the original derivation by substituting for $x^{\phi} : \phi$ the derivation of the premise of $\vee I$. If there is no dependence on the assumption $x^{\phi} : \phi$ in the second premise of $\vee E$, then the result of conversion is the given derivation of that premise $\Delta \Rightarrow \theta$. The $\vee$-conversion is defined similarly when the premise of $\vee$-introduction is $\Gamma \Rightarrow \psi$.

Terminology related to reduction and normalization is transferred to natural deduction. In particular a deduction is *normal* if it does not contain cuts. There is a perfect match between natural deduction and deductive terms.

THEOREM 5.1. *(Curry–Howard isomorphism between terms and natural deductions).*

(a) Every natural deduction d in NJp uniquely defines $\mathcal{T}(d)$ and vice versa: Every term t uniquely defines a natural deduction $\mathcal{D}(t)$.

(b) Cuts in d uniquely correspond to redexes in $\mathcal{T}(d)$, and vice versa.

(c) Every conversion in d uniquely corresponds to a conversion in $\mathcal{T}(d)$, and reduction sequences for d uniquely correspond to reduction sequences for $\mathcal{T}(d)$, and vice versa.

(d) The derivation d is normal iff the term $\mathcal{T}(d)$ is normal.

Proof. Contained in preceding figures. $\dashv$

EXAMPLE 5.2. Consider the following deduction d of a sequent $\alpha \to q \Rightarrow q$, using abbreviations:

$$\alpha \equiv (((p \to q) \to p) \to p),\ \delta \equiv (p \vee (p \to q)) \text{ and } Ax\ \phi \equiv\ \phi \Rightarrow \phi$$

for every formula ϕ.

$$\dfrac{\dfrac{\dfrac{Ax\ \delta \to q \qquad \dfrac{\dfrac{\dfrac{Ax\ \delta \to q \qquad \dfrac{p \Rightarrow p}{p \Rightarrow \delta}}{\delta \to q, p \Rightarrow q}}{\delta \to q \Rightarrow p \to q}}{\delta \to q \Rightarrow \delta}}{\delta \to q \Rightarrow q}}{\Rightarrow (\delta \to q) \to q} \qquad d^q :\ \alpha \to q \Rightarrow \delta \to q}{d :\ \alpha \to q \Rightarrow q}$$

where:

$$\dfrac{\dfrac{Ax\ \alpha \to q \qquad \dfrac{Ax\ \delta \quad \dfrac{p \Rightarrow p}{p \Rightarrow \alpha} \qquad \dfrac{\dfrac{Ax\ (p \to q) \to p \quad Ax\ p \to q}{(p \to q) \to p, p \to q \Rightarrow p}}{p \to q \Rightarrow \alpha}}{\delta \Rightarrow \alpha}}{\alpha \to q, \delta \Rightarrow q}}{d^q :\ \alpha \to q \Rightarrow \delta \to q}$$

Let us compute the term $\mathcal{T}(d)$ assigned to this deduction d and terms assigned to its subdeductions, using variables $x^{\alpha \to q}, y^{\delta \to q}, z^p, u^\delta, v^{p \to q}, w^{(p \to q) \to q}$:

$$\dfrac{\dfrac{\dfrac{y^{\delta \to q} \quad \dfrac{\dfrac{\dfrac{y^{\delta \to q} \quad \dfrac{z^p}{\mathbf{k}_0 z}}{y(\mathbf{k}_0 z)}}{\lambda z.y(\mathbf{k}_0 z)}}{\mathbf{k}_1 \lambda z.y(\mathbf{k}_0 z)}}{y(\mathbf{k}_1 \lambda z.y(\mathbf{k}_0 z))}}{\lambda y.y(\mathbf{k}_1 \lambda z.y(\mathbf{k}_0 z))} \qquad \dfrac{\dfrac{x^{\alpha \to q} \quad \dfrac{u^\delta \quad \dfrac{z^p}{\lambda w.z^p} \quad \dfrac{\dfrac{w^{(p \to q) \to p} \quad v^{p \to q}}{w(v)}}{\lambda w.w(v)}}{D_{z,v}(u, \lambda w.z, \lambda w.w(v))}}{x(D_{z,v}(u, \lambda w.z, \lambda w.w(v)))}}{\lambda u.x(D_{z,v}(u, \lambda w.z, \lambda w.w(v)))}}{(\lambda y.y(\mathbf{k}_1 \lambda z.y(\lambda w.z)))(\lambda u.x(D_{z,v}(u, \lambda w.z, \lambda w.w(v))))}$$

The last $\to E$ and the immediately preceding $\to I$ form a cut. Reduction of this cut leads to the following deduction d_1:

$$\dfrac{d^q :\ \alpha \to q \Rightarrow \delta \to q \qquad \dfrac{\dfrac{\dfrac{d^q :\ \alpha \to q \Rightarrow \delta \to q \quad \dfrac{p \Rightarrow p}{p \Rightarrow \delta}}{\alpha \to q, p \Rightarrow q}}{\alpha \to q \Rightarrow p \to q}}{\alpha \to q \Rightarrow \delta}}{d_1 :\ \alpha \to q \Rightarrow q}$$

Since deduction d^q ends in an introduction rule, both occurrences of $\delta \to q$ derived by d^q give rise to cuts. Reducing the upper cut results in the following deduction of $\alpha \to q, p \Rightarrow q$:

$$\dfrac{Ax\ \alpha \to q \qquad \dfrac{\dfrac{Ax\ p}{p \Rightarrow \delta} \quad \dfrac{p \Rightarrow p}{p \Rightarrow \alpha} \qquad \dfrac{\dfrac{Ax\ (p \to q) \to p \quad Ax\ p \to q}{(p \to q) \to p, p \to q \Rightarrow p}}{p \to q \Rightarrow \alpha}}{p \Rightarrow \alpha}}{d_1^q :\ \alpha \to q, p \Rightarrow q}$$

The $\vee$-introduction with the premise $Ax\ p$ deriving δ in d_1^q is immediately followed by $\vee E$. Reducing this cut results in:

$$\dfrac{Ax\ \alpha \to q \quad \dfrac{p \to p}{p \Rightarrow \alpha}}{d_2^q :\ \alpha \to q, p \Rightarrow q}$$

Thus the original derivation d is reduced to:

$$\dfrac{\dfrac{\dfrac{Ax\,\alpha\to q \qquad \dfrac{Ax\,\delta \quad \dfrac{p\Rightarrow p}{p\Rightarrow\alpha} \quad \dfrac{\dfrac{Ax\,(p\to q)\to p \quad Ax\,p\to q}{(p\to q)\to p, p\to q\Rightarrow p}}{p\to q\Rightarrow\alpha}}{\delta\Rightarrow\alpha}}{\alpha\to q,\delta\Rightarrow q}}{d^q:\ \alpha\to q\Rightarrow\delta\to q} \qquad \dfrac{\dfrac{\dfrac{Ax\,\alpha\to q \quad \dfrac{p\Rightarrow p}{p\Rightarrow\alpha}}{\alpha\to q, p\Rightarrow q}}{\alpha\to q\Rightarrow p\to q}}{\alpha\to q\Rightarrow\delta}}{d_2:\ \alpha\to q\Rightarrow q}$$

Reduce the only cut in d_2 (at the end):

$$\dfrac{Ax\,\alpha\to q \qquad \dfrac{\dfrac{\dfrac{\dfrac{Ax\,\alpha\to q \quad \dfrac{p\Rightarrow p}{p\Rightarrow\alpha}}{\alpha\to q, p\Rightarrow q}}{\alpha\to q\Rightarrow p\to q}}{\alpha\to q\Rightarrow\delta} \quad \dfrac{p\Rightarrow p}{p\Rightarrow\alpha} \quad \dfrac{\dfrac{Ax\,(p\to q)\to q \quad Ax\,p\to q}{(p\to q)\to q, p\to q\Rightarrow p}}{p\to q\Rightarrow\alpha}}{\alpha\to q\Rightarrow\alpha}}{d_3:\ \alpha\to q\Rightarrow q}$$

Reduce the only cut in d_3: $\vee I$, introducing δ followed by $\vee E$:

$$\dfrac{Ax\,\alpha\to q \qquad \dfrac{\dfrac{Ax\,(p\to q)\to q \quad \dfrac{\dfrac{Ax\,\alpha\to q \quad \dfrac{p\Rightarrow p}{p\Rightarrow\alpha}}{\alpha\to q, p\Rightarrow q}}{\alpha\to q\Rightarrow p\to q}}{(p\to q)\to q, \alpha\to q\Rightarrow p}}{\alpha\to q\Rightarrow\alpha}}{d_4:\ \alpha\to q\Rightarrow q}$$

This normal deduction is the same as in Example 2.7..

To get a better feeling of the Curry–Howard isomorphism, let us normalize the term $\mathcal{T}(d)$:

$$(\lambda y.y(\mathbf{k}_1\lambda z.y(\mathbf{k}_0 z)))(\lambda u.x(D_{z,v}(u,\lambda w.z,\lambda w.w(v))))\ conv$$

$$((\lambda u.x(D_{z,v}(u,\lambda w.z,\lambda w.w(v))))(\mathbf{k}_1\lambda z.(\lambda u.x(D_{z,v}(u,\lambda w.z,\lambda w.w(v))))(\mathbf{k}_0 z)))$$

$\equiv \mathcal{T}(d_1)$ red_1

$$(\lambda u.x(D_{z,v}(u,\lambda w.z,\lambda w.w(v))))(\mathbf{k}_1\lambda z.x(D_{z,v}(\mathbf{k}_0 z,\lambda w.z,\lambda w.w(v))))$$

red_1

$$(\lambda u.x(D_{z,v}(u,\lambda w.z,\lambda w.w(v))))(\mathbf{k}_1\lambda z.x(\lambda w.z))\ \equiv \mathcal{T}(d_2)$$

red_1

$$x(D_{z,v}(\mathbf{k}_1\lambda z.x(\lambda w.z),\lambda w.z,\lambda w.w(v)))$$

$\equiv \mathcal{T}(d_3)$ red_1

$$x(\lambda w.w(\lambda z.x(\lambda w.z)))\equiv \mathcal{T}(d_4)\equiv|\mathcal{T}(d)|$$

5.3. Normalization

Let us measure complexity of a formula by its *length*, that is, the number of occurrences of logical connectives:

$$lth(p) = 0; \qquad lth(\phi\&\psi) = lth(\phi \vee \psi) = lth(\phi \to \psi) := lth(\phi) + lth(\psi) + 1$$

The complexity or *cutrank* of a cut in a deduction is the length of its cut formula. In the language of deductive terms:

$$cutrank((\lambda x^{\phi}.t^{\psi})u^{\phi}) = cutrank(\mathbf{p}_i\mathbf{p}(t^{\phi}, s^{\psi})) = cutrank(D_{x^{\phi},y^{\psi}}(\mathbf{k}_i t, s_0, s_1))$$

$$:= lth(\phi) + lth(\psi) + 1$$

Let $maxrank(t)$ be the maximal complexity of redeces in a term t (and 0 if t is normal).

LEMMA 5.2. *(a) If t, s are deductive terms, $t \not\equiv x^{\phi}$, and $t[x^{\phi}/s^{\phi}]$ is a redex, then either t is a redex (and $cutrank(t) = cutrank(t[x/s])$) or one of the following conditions is satisfied:*

$$\begin{array}{ccccc} t \equiv x(t') & & t \equiv \mathbf{p}_i x & & t \equiv D(x, t_0, t_1) \\ s \equiv \lambda y.s' & or & s \equiv \mathbf{p}(s_0, s_1) & or & s \equiv \mathbf{k}_i s' \end{array} \tag{5.6}$$

and $cutrank(t[x/s]) = lth(\phi)$

(b) If t^{ϕ} conv t' and $cutrank(t) > cutrank(s)$ for every proper subterm s of t, then $maxrank(t) = cutrank(t) > maxrank(t')$.

Proof. Part (a) says that really new redeces in a term can arise after a substitution only where an elimination rule was applied to a variable substituted by an introduction term. Indeed it is easy to see by inspection that every other non redex goes into a non redex. A complete proof is done by induction on the construction of t.

To prove (b), note that:

$$maxrank(t) = cutrank(t) > maxrank(s) \tag{5.7}$$

for every proper subterm s by the assumption, and consider possible cases. If $t \equiv \mathbf{p}_i\mathbf{p}(t_0, t_1)$ *conv* t_i, then $maxrank(t) > maxrank(t_i)$ by (5.7). If $t \equiv (\lambda x^{\phi}.t_0)(s)$ *conv* $t_0[x/s] \equiv t'$, then by Part (a) every redex in t' either has the same cutrank as some redex in t_0 [which is less than $cutrank(t)$ by the assumption] or has cutrank $length(\phi) < cutrank(t)$.

If $t \equiv D_{x_0,x_1}(\mathbf{k}_i t, s_0, s_1))$ *conv* $s_i[x_i/t]$, the argument is similar. ⊣

THEOREM 5.2. *(normalization theorem). (a) Every deductive term t can be normalized.*

(b) Every natural deduction d can be normalized.

Proof. Part (b) follows from Part (a) by the Curry–Howard isomorphism. For Part (a) we use a main induction on $n = maxrank(t)$ with a subinduction on m, the number of redeces of cutrank n.

The induction base is obvious for both inductions. For the induction step on m, choose in t the rightmost redex ρ of the cutrank n and convert it into its reductum ρ'. Since ρ is the rightmost, it does not have proper subterms of cutrank n. By Lemma 5.3.(b) $maxrank(\rho) = n > maxrank(\rho')$. Write $t \equiv t'[y/\rho]$ to indicate the unique occurrence of ρ in t: The variable y has exactly one occurrence in t', term t' has exactly $m-1$ redeces of cutrank n, and

$$t \equiv t'[y^{\phi}/\rho^{\phi}] \; conv \; t'[y/\rho']$$

Applying Lemma 5.3.(a) to $t'[y/\rho']$, new redeces have cutranks equal to $lth(\phi) < maxrank(\rho') < n$, and old redeces preserve their cutranks. Since the redex ρ of cutrank n disappeared, the m decreased by one, and the induction step is proved. ⊣

5.4. Consequences of Normalization

The *principal formula* of an elimination rule is the succedent formula explicitly shown in the rule and containing eliminated connective: $\alpha\&\beta$ in $\&E$, and so on. The *principal premise* contains the principal formula.

A *main branch* of a deduction is a branch ending in the final sequent and containing principal premises of elimination rules with conclusions in the main branch. Hence the main branch of a deduction ending in an introduction rule contains only the final sequent. In any case the main branch is the leftmost branch up to the lowermost introduction rule or axiom.

THEOREM 5.3. *(properties of normal deductions). Let $d : \Gamma \Rightarrow \gamma$ be a normal deduction in NJp.*

(a) If d ends in an elimination rule, then the main branch contains only elimination rules, begins with an axiom, and every sequent in it is of the form $\Gamma' \Rightarrow \alpha$, where $\Gamma' \subset \Gamma$ and α is some formula.

(a1) In particular the axiom at the top of the main branch is of the form $\alpha \Rightarrow \alpha$ where $\alpha \in \Gamma$.

(b) If $\Gamma = \emptyset$, then d ends in an introduction rule

Proof. Part (a): If d ends in an elimination rule, then the main branch does not contain an introduction rule: Conclusion of such a rule would be a cut. Now Part (a) is proved by induction on the number of rules in the main branch using an *observation*: An antecedent of the principal premise of an elimination rule is contained in the antecedent of the conclusion. Part (a1) immediately follows from (a).

Part (b): Otherwise the main branch of d cannot begin with an axiom by (a). ⊣

THEOREM 5.4. *(disjunction property, Harrop's theorem).*

(a) If $\vdash \alpha_0 \vee \alpha_1$, *then* $\vdash \alpha_i$ *for some* $i = 0, 1$.

(b) If $\neg\beta \vdash \alpha_0 \vee \alpha_1$, *then* $\neg\beta \vdash \alpha_i$ *for some* $i = 0, 1$.

Proof. Part (a) follows from Theorems 5.3. and 5.4.(b). For Part (b) consider the last (lowermost) rule of a given normal deduction of the sequent in question. If it is an introduction, we are done, as in Part (a). If it is an elimination, consider the axiom and the very first (uppermost) rule in the main branch. The axiom is $\neg\beta \Rightarrow \neg\beta$ by Theorem 5.4.(a). Hence the first rule is $\rightarrow$-elimination (recall that $\neg\beta = \beta \rightarrow \bot$):

$$\frac{\beta \rightarrow \bot \Rightarrow \beta \rightarrow \bot \quad \beta \rightarrow \bot \Rightarrow \beta}{\beta \rightarrow \bot \rightarrow \bot}$$

Conclusion of this rule implies $\neg\beta \Rightarrow \gamma$ for any γ. ⊣

COROLLARY 5.1. *The law of the excluded middle and the law of double negation are not derivable in NJp:*

$$\nvdash p \vee \neg p, \qquad \nvdash \neg\neg p \rightarrow p$$

Proof. By disjunction property $\vdash p \vee \neg p$ implies that one of $p, \neg p$ is derivable, but none of these is even a tautology. If $\vdash \neg\neg p \rightarrow p$, then substituting $p := q \vee \neg q$ and using $\neg\neg(q \vee \neg q)$ (Example 2.4.), we obtain $\vdash q \vee \neg q$. ⊣

Chapter 6

Coherence Theorem

6.1. Structure of Normal Deduction

An occurrence of a subformula is *positive* in a formula if it is in the premise of an even number (maybe 0) of occurrences of implication. An occurrence is *strictly positive* if it is not in the premise of any implication. An occurrence is *negative* if it is not positive, that is, it is inside an odd number of premises of implication. The sign of an occurrence in a sequent $\Gamma \Rightarrow \alpha$ is the same as in the formula $\&\Gamma \rightarrow \alpha$.

THEOREM 6.1. *(subformula property). Let $d : \Gamma \Rightarrow \gamma$ be a normal $\vee$-free deduction.*

(a) If d ends in an elimination rule, then the main branch begins with an axiom $\alpha \Rightarrow \alpha$ for $\alpha \in \Gamma$ and all succedents in the main branch are strictly positive subformulas of α (and hence of Γ).

(b) All formulas in d are subformulas of the last sequent.

Proof. Part (a) is proved by an easy induction on the length of d. The induction base and the case when d ends in an introduction rule are trivial. If d ends in an elimination rule L, the major premise of L takes the form $\Gamma' \Rightarrow \gamma'$, with $\Gamma' \subset \Gamma$ and γ' strictly positive in α by IH. Since the succedent in the conclusion of $\&E, \rightarrow E$ is strictly positive in the major formula γ', this succedent is strictly positive in α as required.

Part(b): Induction on the deduction d. The induction base (axiom) is trivial. In the induction step, consider cases depending of the last rule L:

Case 1. The L is an introduction rule. Then all formulas in premises are subformulas of the conclusion, and the subformula property follows from IH.

Case 2. The L is an elimination rule, say:

$$\frac{\Gamma \Rightarrow \alpha \rightarrow \beta \quad \Delta \Rightarrow \alpha}{[\Gamma, \Delta] \Rightarrow \beta}$$

By part (a) $\alpha \to \beta$ is a subformula of the last sequent. By IH all subformulas in subdeductions are subformulas of $\Gamma, \alpha \to \beta, \Delta, \alpha$, and hence of the last sequent. ⊣

6.2. η-reduction

For applications to category theory, we require a stronger reduction relation than β-reduction. The *η-conversion* for deductive terms corresponding to deductions in the language $\{\&, \to\}$ is defined as follows:

$$\mathbf{p}(\mathbf{p}_0(t), \mathbf{p}_1(t)) \ conv\ t,$$

$$\lambda x.(tx) \ conv\ t \quad \text{provided } x \notin FV(t).$$

Corresponding conversions for deductions are as follows:

$$\dfrac{\dfrac{d : \Gamma \Rightarrow \phi_0 \& \phi_1}{\Gamma \Rightarrow \phi_0} \quad \dfrac{d : \Gamma \Rightarrow \phi_0 \& \phi_1}{\Gamma \Rightarrow \phi_1}}{\Gamma \Rightarrow \phi_0 \& \phi_1} \qquad conv \quad d : \Gamma \Rightarrow \phi_0 \& \phi_1$$

$$\dfrac{\dfrac{d :\ \Gamma \Rightarrow \alpha \to \beta \quad \alpha \Rightarrow \alpha}{\Gamma, \alpha \Rightarrow \beta}}{\Gamma \Rightarrow \alpha \to \beta} \qquad conv \quad d :\ \Gamma \Rightarrow \alpha \to \beta$$

Hence the Curry–Howard isomorphism (Theorem 5.2.) is preserved.

The $\beta\eta$-conversion is a combination of these conversions and (5.1),(5.2). The η-reduction, $\beta\eta$-reduction, and corresponding normal forms $|t|_\eta$, $|t|_{\beta\eta}$ are defined as for β-conversion. These normal forms are unique, but we shall not prove it here.

LEMMA 6.1. *(a) Every η-reduction sequence terminates.*

(b) Every deductive term and every deduction has a $\beta\eta$-normal form.

Proof. Part (a): Every η-conversion reduces the size of the term.

Part (b): A β-normal form $|t|_\beta$ exists by Theorem 13.3., and its η-normal form [see Part (a)] is $\beta\eta$-normal, since η-conversions preserve β-normal form. ⊣

6.3. Coherence Theorem

In this section we consider $NJp_{\to}$-deductions of implicative formulas and corresponding deductive terms modulo $\beta\eta$-conversion: The $d = d'$ stands for $|d|_{\beta\eta} = |d'|_{\beta\eta}$ and similarly for $t = t'$.

A sequent is *balanced* if every propositional variable occurs there at most twice and at most once with a given sign (positively or negatively; see Section 6.1.).

Example. $p \to (q \to r) \Rightarrow q \to (p \to r)$ and $(p \to q) \to r \Rightarrow q \to r$ are balanced, but $p, p \to p \Rightarrow p$ is not.

We prove that a balanced sequent has unique deduction up to $\beta\eta$-equality. For non-balanced sequents that is false: The sequent $p, p \to p \Rightarrow p$ has infinitely many different normal proofs:

$$\frac{p \to p \Rightarrow p \to p \quad p \Rightarrow p}{d_1 : \ p, p \to p \Rightarrow p} \qquad \frac{p \to p \Rightarrow p \to p \quad d_n : \ p, p \to p \Rightarrow p}{d_{n+1} : \ p, p \to p \Rightarrow p}$$

The d_n can be described as a "component" of the unique proof of the balanced sequent $p_1, p_1 \to p_2, \ldots, p_n \to p_{n+1} \Rightarrow p_{n+1}$ obtained by identifying all variables with p.

Note. Formulas of $NJp_{\to}$ as objects and the normal NJp-deductions as morphisms form a $\to$-part of a *Cartesian closed category.* Theorem 6.3. below shows that a morphism $d : \alpha \Rightarrow \beta$ with a balanced $\alpha \to \beta$ is unique. In fact Theorem 6.3. extends to the language $\{\&, \to\}$ ([1], [16]). Abbreviation: $(\alpha_1 \ldots \alpha_n \to \beta) := (\alpha_1 \to \ldots \to (\alpha_n \to \beta) \ldots)$.

The next Lemma shows that some of the redundant assumptions are pruned by normalization. Recall that notation $\delta^0, \Gamma \Rightarrow \alpha$ means that δ may be present or absent.

LEMMA 6.2. *(pruning lemma). (a) Assume that Σ, α are implicative formulas, propositional variable q does not occur positively in $\Sigma \Rightarrow \alpha$, and a deduction $d : (\Delta \to q)^0, \Sigma \Rightarrow \alpha$ is normal; then $d : \Sigma \Rightarrow \alpha$.*

(b) If $NJp_{\to} \vdash (\alpha_1, \ldots, \alpha_n \to q)$, then one of α_i contains q positively.

Proof. For Part (a) use induction on d. Induction base and the case when d ends in an introduction rule are obvious. Let d end in an $\to E$. Consider the main branch of d, which by Theorem 6.1. begins with an axiom $\mathcal{A} \Rightarrow \mathcal{A}$ for $\mathcal{A} \equiv (\alpha_1 \ldots \alpha_n \to \alpha)$, since α is strictly positive in $\mathcal{A}$:

$$\begin{array}{c}
\mathcal{A} \Rightarrow (\alpha_1 \ldots \alpha_n \to \alpha) \\
\cdots \\
\dfrac{[\mathcal{A}, (\Delta \to q)^*, \Gamma_1, \ldots, \Gamma_{i-1}] \Rightarrow (\alpha_i \ldots \alpha_n \to \alpha) \quad (\Delta \to q)', \Gamma_i \Rightarrow \alpha_i}{[\mathcal{A}, (\Delta \to q)^{**}, \Gamma_1, \ldots, \Gamma_i] \Rightarrow (\alpha_{i+1} \ldots \alpha_n \to \alpha)} \\
\cdots \\
[\mathcal{A}, (\Delta \to q)^0, \Gamma_1, \ldots, \Gamma_n] \Rightarrow \alpha
\end{array}$$

Superscripts attached to the assumption $\Delta \to q$ indicate that it may be absent from some of the sequents. Since q is not positive in α, the formula $\mathcal{A}$ in the axiom of the main branch is distinct from $(\Delta \to q)$. Since $\mathcal{A}$ occurs in the antecedent of the last sequent, q is not negative in $\mathcal{A}$, and hence it is not positive in α_i, since $(\alpha_i, \ldots, \alpha_n \to \alpha)$ is strictly positive in $\mathcal{A}$. All other formulas in the minor premises $(\Delta \to q)', \Gamma_i \Rightarrow \alpha_i$ have the same sign in the last sequent. Hence IH is applicable to all minor premises, and $(\Delta \to q)$ is not present in the antecedent.

Part (b): Assign $q := 0$, $p := 1$ for all $p \not\equiv q$ and compute by truth tables. If all α_i are of the form $\Pi \to p$, and hence true, then $\alpha_1, \ldots, \alpha_n \to q$ is false under our assignment. Thus it is not even a tautology. Alternatively, apply (a). $\dashv$

THEOREM 6.2. *(coherence theorem). (a) Let $d, d' : \Rightarrow \alpha$ for a balanced implicative formula α; then $d = d'$.*

(b)Let $[\Gamma, \Gamma'] \Rightarrow \alpha$ be balanced, $d : \Gamma \Rightarrow \alpha$, $d' : \Gamma' \Rightarrow \alpha$; then $d = d'$.

Proof. Part (a) follows from Part (b), which claims that Γ and Γ' are pruned during normalization into one and the same set of formulas. Since $[\Gamma, \Gamma']$ is balanced, each of Γ, Γ' is balanced. To prove Part (b), we apply induction on the length of $[\Gamma, \Gamma'] \Rightarrow \alpha$. Assume $d : \Gamma \Rightarrow t : \alpha$, $d : \Gamma' \Rightarrow t' : \alpha$ and recall that $d = d'$ iff $t = t'$.

Case 1. The $\alpha \equiv (\beta \rightarrow \gamma)$; then $[(\beta, \Gamma), \Gamma'] \Rightarrow \gamma \equiv [(\beta, \Gamma), (\beta, \Gamma')] \Rightarrow \gamma$ is balanced, and IH is applicable to sequents obtained by applying $\rightarrow E$-rule with the minor premise $\beta \Rightarrow \beta$ to d, d'. This corresponds to applying a new variable x^β to deductive terms $\mathcal{T}(d), \mathcal{T}(d')$. We have $(\mathcal{T}(d), x^\beta) = (\mathcal{T}(d'), x^\beta)$; hence $\mathcal{T}(d) = \lambda x^\beta(\mathcal{T}(d), x^\beta) = \lambda x^\beta(\mathcal{T}(d'), x^\beta) = \mathcal{T}(d')$ and $d = d'$.

Case 2. The α is a propositional variable; then each of the $\beta\eta$-normal forms $|d|, |d'|$ is an axiom or ends in $\rightarrow E$.

Case 2.1. The $|d|$ is an axiom $\alpha \Rightarrow \alpha$; then no member of $[\Gamma, \Gamma']$ different from α contains α positively, and by the Lemma 6.3. (a), we have $|d'| : \alpha \Rightarrow \alpha$; that is, $d = d'$.

Case 2.2. Both $|d|$ and $|d'|$ end in $\rightarrow E$. Consider the main branch of each of these deductions. Since α is strictly positive in the axiom formula of the main branch (Theorem 6.1.), and $[\Gamma, \Gamma'] \Rightarrow \alpha$ is balanced, this axiom formula $\mathcal{A} \equiv \alpha_1 \ldots \alpha_n \rightarrow \alpha$ is one and the same in $|d|$ and $|d'|$ and the number of $\rightarrow E$-inferences in the main branch is the same:

$$\cfrac{\begin{array}{c}\mathcal{A} \Rightarrow \alpha_1 \ldots \alpha_n \rightarrow \alpha \\ \vdots \\ [\mathcal{A}, \Gamma_1, \ldots, \Gamma_{i-1}] \Rightarrow \alpha_i \ldots \alpha_n \rightarrow \alpha \quad d_i : \Gamma_i \Rightarrow \alpha_i\end{array}}{\begin{array}{c}[\mathcal{A}, \Gamma_1, \ldots, \Gamma_i] \Rightarrow \alpha_{i+1} \ldots \alpha_n \rightarrow \alpha \\ \vdots \\ |d| : [\mathcal{A}, \Gamma_1, \ldots, \Gamma_n] \Rightarrow \alpha\end{array}}$$

$$\cfrac{\begin{array}{c}\mathcal{A} \Rightarrow \alpha_1 \ldots \alpha_n \rightarrow \alpha \\ \vdots \\ [\mathcal{A}, \Gamma'_1, \ldots, \Gamma'_{i-1}] \Rightarrow \alpha_i \ldots \alpha_n \rightarrow \alpha \quad d'_i : \Gamma'_i \Rightarrow \alpha_i\end{array}}{\begin{array}{c}[\mathcal{A}, \Gamma'_1, \ldots, \Gamma'_i] \Rightarrow \alpha_{i+1} \ldots \alpha_n \rightarrow \alpha \\ \vdots \\ |d'| : [\mathcal{A}, \Gamma'_1, \ldots, \Gamma'_n] \Rightarrow \alpha\end{array}}$$

The only positive occurrence of the propositional variable α in a balanced sequent:

$$[\mathcal{A}, \Gamma_1, \ldots, \Gamma_n] \Rightarrow \alpha$$

is the succedent, and the same is true for $[\mathcal{A}, \Gamma'_1, \ldots, \Gamma'_n] \Rightarrow \alpha$. In particular α is not negative in $\Gamma_1, \ldots, \Gamma_n$, $\Gamma'_1, \ldots, \Gamma'_n$ and in $\mathcal{A}$; hence α is not positive in $\alpha_1, \ldots, \alpha_n$. By the Lemma 6.3. (a) the formula $\mathcal{A}$ is not a member of $\Gamma_1, \ldots, \Gamma_n, \Gamma'_1, \ldots, \Gamma'_n$; hence each of $[\Gamma_i, \Gamma'_i] \Rightarrow \alpha_i$ is balanced. Indeed compare the following:

$$[\Gamma_i, \Gamma'_i] \Rightarrow \alpha_i \quad \text{and} \quad (\alpha_1, \ldots, \alpha_i, \ldots, \alpha_n \rightarrow \alpha), [\Gamma, \Gamma'] \Rightarrow \alpha.$$

Every occurrence in $[\Gamma_i, \Gamma_i']$ is uniquely matched with an occurrence of the same sign in:

$$[\Gamma, \Gamma'] \equiv [(\Gamma_1, \ldots, \Gamma_i, \ldots, \Gamma_n), (\Gamma_1', \ldots, \Gamma_i', \ldots, \Gamma_n')].$$

Every occurrence in α_i is uniquely matched with an occurrence of the same sign generated by an occurrence of α_i in $\mathcal{A} \equiv (\alpha_1 \ldots, \alpha_i, \ldots, \alpha_n \to \alpha)$. Applying IH to deductions of $\Gamma_i \Rightarrow A_i$ and $\Gamma_i' \Rightarrow \alpha_i$ yields $d_i = d_i'$; hence $|d| = |d'|$ as required. ⊣

Chapter 7

Kripke Models

Recall that a model for the classical propositional calculus is simply an assignment of the truth values *true* (1) and *false* (0) to the propositional variables. This reflects the state of the world: Some atomic statements are true, and some are false.

The semantics for intuitionistic logic described in the following reflects a more dynamic approach: Our current knowledge about the truth of statements can improve. Some statements whose truth status was previously indeterminate can be established as true. The value *true* corresponds to firmly established truth that is preserved with the advancement of knowledge, and the value *false* corresponds to "not yet true". To refute a formula ϕ, that is, to establish $\neg\phi$ at a stage w, it is necessary that $V(\phi, w') = false$ for all future stages w'.

We show that semantical validity is equivalent to derivability in intuitionistic propositional logic.

The various stages of knowledge, or *worlds* as these are called, are simply truth value assignments for propositional variables. An important feature of this approach is an accessibility relation R between worlds: $R\,w\,w'$ is read as "w' is accessible from w" and it is interpreted as "w' is more advanced than w or $w' = w$ ". Such interpretation requires R to be reflexive and transitive:

$$Rww \quad \text{and} \quad Rww' \& Rw'w'' \to Rww'' \text{ for all } w$$

This motivates the following Definition.

DEFINITION 7.1. *(propositional intuitionistic model). A propositional intuitionistic* model *is an ordered triple* $\langle W, R, V\rangle$, *where* W *is a non-empty set,* R *is a binary* reflexive *and* transitive *relation on* W, *and* V *is a function assigning a truth value* $0, 1$ *to each propositional variable* p *in each* $w \in W$*:*

$$V(p, w) \in \{0, 1\}$$

V *is assumed to be* monotone *with respect to* R*:*

$$V(p, w) = 1 \textit{ and } R\,w\,w' \textit{ implies } V(p, w') = 1$$

Elements w of the set W are called worlds, *R is an* accessibility relation, *and V is a* valuation function. *A pair $\langle W, R\rangle$ is called a* (intuitionistic Kripke) frame.

The following definition was introduced by Kripke for intuitionistic logic; it is connected to his previous treatment of modal logic. (It is easy to recognize its common features with the notion of forcing introduced by Cohen.)

DEFINITION 7.2. *A truth value $V(\varphi, w) \in \{0,1\}$ for arbitrary propositional formula φ and a world $w \in W$ in a model $\langle W, R, V\rangle$ is defined by recursion on φ:*

p : *$V(p, w)$ for a propositional variable p already defined.*

$\bot : V(\bot, w) := 0$

$\varphi \& \psi : V(\varphi \& \psi, w) := V(\varphi, w) \& V(\psi, w) = \min(V(\varphi, w), V(\psi, w))$

$\varphi \vee \psi : V(\varphi \vee \psi, w) := V(\varphi, w) \vee V(\psi, w) = \max(V(\varphi, w), V(\psi, w))$

$\varphi \to \psi : V(\varphi \to \psi, w) = 1$ *iff* $V(\varphi, w') = 1$ *implies* $V(\psi, w') = 1$ *for all* $w' \in W$ *such that* $R\,w\,w'$.

As a consequence, we have the following:

$\neg\varphi : V(\neg\varphi, w) = 1$ iff $V(\varphi, w') = 0$ for all $w' \in W$ such that $R\,w\,w'$.

In other words, conjunction and disjunction behave classically (in a boolean way) in each world. The $\neg\varphi$ is true at world w iff φ will always be false. The $\varphi \to \psi$ is true in w iff the truth of φ implies the truth of ψ in every accessible world. In terms of stages of knowledge $\neg\varphi$ says that no possible advancement of knowledge will justify φ. The $\varphi \to \psi$ says that whenever φ is justified, ψ will also be justified.

Instead of $V(\varphi, w) = 1$, we sometimes write $w \models \varphi$.

Note that $w \models \neg\varphi$ means $(\forall w' : R\,w\,w')(\text{not } w' \models \varphi)$.

The formula φ is *true at the world* w iff $V(\varphi, w) = 1$, and φ is *valid* in a model $\langle W, R, V\rangle$ iff it is true at every world $w \in W$. The latter relations are denoted by $M, w \models \varphi$ and $M \models \varphi$. Finally φ is *valid* (written $\models \varphi$) iff it is valid in all (propositional intuitionistic) models.

EXAMPLE 7.1. . Let $W = \{w_0, w_1\}$, and let the relation R be given by $R\,w_0\,w_0$, $R\,w_0\,w_1$, $R\,w_1\,w_1$. Finally let V be given by $V(p, w_1) = 1$ and V be false for any other propositional letter. Graphically we describe this by:

The w_1 is over w_0 and connected to w_0, since $R\,w_0\,w_1$. Let us compute the value $V(p \vee \neg p, w_0)$:

$V(p \vee \neg p, w_0) = V(p, w_0) \vee V(\neg p, w_0) = 0 \vee V(\neg p, w_0)$.
$V(\neg p, w_0) = 0$ since $R\, w_0\, w_1$ and $V(p, w_1) = 1$.
So:

$$V(p \vee \neg p, w_0) = 0.$$

That is, the law of the excluded middle is refuted in our model. Verify that the same model refutes the law of double negation:

$$\neg\neg p \rightarrow p$$

With Theorem 7.1. below this example again shows that $p \vee \neg p$ and $\neg\neg p \rightarrow p$ are not derivable in NJp.

EXAMPLE 7.2. The model

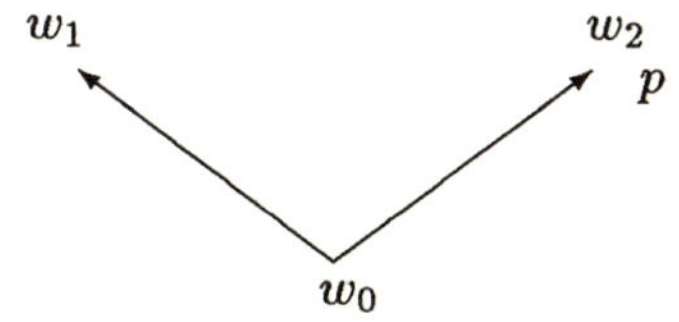

that is, $V(p, w_0) = V(p, w_1) = 0$; $V(p, w_2) = 1$ and $R\, w_0\, w_1, R\, w_0\, w_2$, refutes the principle of the weak excluded middle $\neg p \vee \neg\neg p$. Indeed $V(\neg p, w_1) = 1$, since w_1 is the only w' such that $R\, w_1\, w'$. $V(\neg p, w_0) = 0$, since $R\, w_0\, w_2$ and $V(p, w_2) = 1$. $V(\neg\neg p, w_0) = 0$, given that $R\, w_0\, w_1$, and $V(\neg p, w_1) = 1$; therefore:

$$V(\neg p \vee \neg\neg p, w_0) = 0 \vee 0 = 0$$

Let us prove that truth is monotonic with respect to R.

LEMMA 7.1. *(monotonicity lemma). Let $\langle W, R, V \rangle$ be a model; then for any $w, w' \in W$ and formula φ:*

$$R\, w\, w' \textit{ and } V(\varphi, w) = 1 \textit{ imply } V(\varphi, w') = 1. \tag{7.1}$$

Proof. Induction on (the construction of) φ. We restate (7.1) as:

$$R(w, w') \rightarrow (V(\varphi, w) \rightarrow V(\varphi, w'))$$

Basis: φ is a propositional variable; then (7.1) is included in the definition of a model.

Induction step: We assume $R\, w\, w'$ and consider cases depending on the main connective of φ.

$\varphi = \psi \& \theta$.

Then

$$V(\varphi, w) = V(\psi, w) \& V(\theta, w)$$

By the IH:

$$V(\psi, w) = 1 \rightarrow V(\psi, w') = 1; \quad V(\theta, w) = 1 \rightarrow V(\theta, w') = 1;$$

Therefore:

$$V(\varphi, w) = 1 \rightarrow V(\psi, w')\&V(\theta, w') = V(\varphi, w') = 1$$

$\varphi = \psi \rightarrow \theta$.
Assume

$$V(\varphi, w) = 1 \tag{7.2}$$

To prove $V(\varphi, w') = 1$, assume $R w' w''$ and $V(\psi, w'') = 1$. Since R is transitive (and we assume $R w w'$), we have $R w w''$. Then by (7.2) we have $V(\theta, w'') = 1$, so $V(\varphi, w'') = 1$ as required.

The other cases are similar. ⊣

7.1. Soundness of the System NJp

The value of a sequent $\Gamma \Rightarrow \beta$ in a model is defined exactly as for the corresponding formula $\&\Gamma \rightarrow \beta$:
$V(\alpha_1, \ldots, \alpha_m \Rightarrow \beta, w) = 1$ iff for any w' such that Rww'
$V(\alpha_1, w') = \ldots V(\alpha_m, w') = 1$ implies $V(\beta, w') = 1$.

THEOREM 7.1. *All the rules of NJp are sound: If all the premises are true in a world w of a Kripke model, then the conclusion is also true in w.*

Proof. By inspection of the rules; consider only two of these.
1. $\&I$.

$$\frac{\Gamma \Rightarrow \phi \quad \Delta \Rightarrow \psi}{[\Gamma, \Delta] \Rightarrow \phi\&\psi}$$

Assume that:

$$V((\Gamma \Rightarrow \phi), w) = V(\Delta \Rightarrow \psi), w) = 1 \tag{7.3}$$

as well as $V(\&\Gamma, w') = 1$, $V(\&\Delta, w') = 1$, $R(w, w')$. By monotonicity, $V((\Gamma \Rightarrow \phi), w') = V(\Delta \Rightarrow \psi), w') = 1$. Now (7.3) implies $V(\phi, w') = V(\psi, w') = 1$; hence $V(\phi\&\psi, w') = 1$, as required.

$\rightarrow I$

$$\frac{\phi^0, \Gamma \Rightarrow \psi}{\Gamma \Rightarrow \phi \rightarrow \psi}$$

Assume $V(\phi^0, \Gamma \Rightarrow \psi, w) = 1$ and $V(\&\Gamma, w') = 1, Rww'$. To prove $V(\Gamma \Rightarrow \phi \rightarrow \psi), w') = 1$, we assume $Rw'w''$ and $V(\phi, w'') = 1$. We must establish $V(\psi, w'') = 1$. By transitivity of the relation R, we have Rww''; and by monotonicity, $V(\phi^0, \Gamma \Rightarrow \psi, w'') = 1$ and $V(\&\Gamma, w'') = 1$. Since R is reflexive, we have $Rw''w''$; by the truth condition for the sequent, we have $V(\psi, w'') = 1$ as required. ⊣

7.2. Pointed Frames, Partial Orders

In a subclass of frames and models, an "actual world" is distinguished.

DEFINITION 7.3. *A* pointed frame *is a triple* $< G, W, R >$, *where* $< W, R >$ *is a frame,* $G \in W$ *and* RGw *for all* $w \in W$. *A* pointed model *is a tuple* $M =< G, W, R, V >$, *where* $< G, W, R >$ *is a pointed frame and* V *is a valuation on* $< W, R >$. *Truth in* M *is truth in the world* G*:*

$$M \models \varphi \textit{ iff } V(G, \varphi) = 1.$$

The next Lemma shows that the truth in a world w depends only on truth values in worlds accessible from w.

LEMMA 7.2. *A formula is valid iff it is true in all pointed models.*

Proof. The implication in one direction is obvious. For other direction, assume that φ is not valid, that is, $M =< W, R, V > \not\models \varphi$ for some M. Then $V(G, \varphi) = 0$ for some $G \in W$. Consider the pointed restriction $M' =< G, W', R, V >$ of M to worlds accessible from G:

$$W' = \{w' \in w : \ RGw\}.$$

By induction on ψ we easily prove that its value in M and M' is always the same:

$$M, w \models \psi \text{ iff } \qquad M', w \models \psi \quad \text{for all } w \in W'. \tag{7.4}$$

The transitivity of R ensures that all necessary worlds from W are present in W' when ψ is an implication or negation. From (7.4) it follows that $M', G \not\models \varphi$ as required. ⊣

DEFINITION 7.4. *A binary relation* R *on a set* W *is a* partial order *iff* R *is reflexive, transitive, and antisymmetric:* $Rww' \& Rw'w \rightarrow w = w'$.

THEOREM 7.2. *A formula* φ *is valid iff it is true in all pointed models* $< G, W, R >$ *partially ordered by* R.

Proof. Set $w \sim w'$ iff Rww' and $Rw'w$. The reflexive transitive relation R may fail to be a partial order due only to failure of antisymmetry: $w \neq w'$ for some $w \sim w'$. However such worlds are indistinguishable by the values of V, since monotonicity implies that:

$$w \sim w' \;\rightarrow\; V(\varphi, w) = V(\varphi, w')$$

for every formula φ.

For the non-trivial part of Theorem, in a pointed model $< G, W, R >$ in which all worlds are accessible from G, identify indistinguishable worlds. More precisely, let $\hat{W}$ be the set of equivalence classes and let $\hat{R}$ be the corresponding accessibility relation:

$$\hat{w} := \{w' : w \sim w'\}; \qquad \hat{W} := \{\hat{w} : w \in W\}; \qquad \hat{R}\hat{w}\hat{w}' \equiv Rww'.$$

Then $\hat{R}$ is a partial order, and the following valuation:

$$\hat{V}(p, \hat{w}) = V(p, w) \tag{7.5}$$

is well-defined and monotonic. It is easy to prove by induction on φ that (7.5) extends to all formulas: $\hat{V}(\varphi, \hat{w}) := V(\varphi, w)$. ⊣

7.3. Frame Conditions

Let us illustrate the use of frame properties for characterizing superintuitionistic logics, that is, extensions of intuitionistic logic.

LEMMA 7.3. *Let $F =< G, W, R >$ be a pointed frame and let every $w \in W$ be accessible from G. Then the law of the excluded middle $p \vee \neg p$ is valid in F iff R is total: Rww' for all $w, w' \in W$. If R is a partial order, then $p \vee \neg p$ is valid iff W is a singleton $\{G\}$.*

Proof. Let us first assume R is total and establish $G \models p \vee \neg p$. If $G \models \neg p$, we are done. Otherwise $w \models p$ for some $w \in W$. Since R is total, we have RwG, and by monotonicity, $G \models p$ as required.

Now assume that R is not total, that is, $\neg Rw_0G$ for some $w_0 \in W$. Define a truth value assignment V as follows:

$$V(p, w) = 1 \text{ iff } Rw_0w$$

V is monotone and $V(p, G) = 0$ by definition of V. On the other hand, $V(\neg p, G) = 0$, since RGw_0 and $V(p, w_0) = 1$. ⊣

EXERCISE 7.1. Find frame conditions for $(p \to q) \vee (q \to p)$ and $\neg p \vee \neg\neg p$.

EXERCISE 7.2.

Prove that a formula $p \vee \neg p$ is valid in a frame $< W, R >$ iff R is symmetric: Rww' implies $Rw'w$ for all $w, w' \in W$.

Chapter 8

Gentzen-type Propositional System LJpm

We prove natural deduction system NJp sound and complete for Kripke models. However it is convenient first to introduce another propositional system LJpm that is equivalent to NJp and more suitable for proof-search. Here we work with *multiple-succedent sequents*, that is, expressions of the form:

$$\phi_1, \ldots, \phi_m \Rightarrow \psi_1 \ldots, \psi_n \qquad (m \geq 0, n \geq 1) \tag{8.1}$$

where ϕ_i, ψ_j are formulas.

The translation of sequents into formulas is given by:

$$(\phi_1, \ldots, \phi_n \Rightarrow \psi_1 \ldots, \psi_n)^f = (\phi_1 \& \ldots \& \phi_n \supset \psi_1 \vee \ldots \vee \psi_n) \tag{8.2}$$

$$= \&_i \phi_i \supset \bigvee_j \psi_j$$

The formula corresponding to a sequent $\Gamma \Rightarrow \Delta$ is written as $\&\Gamma \Rightarrow \vee\Delta$. In particular the sequent $\Rightarrow \psi$ corresponds to the formula ψ, and the sequent $\phi_1, \ldots, \phi_m \Rightarrow$ is translated as $\neg\&_i\phi_i$; it is read "$\phi_1, \ldots, \phi_n$ are contradictory". The left-hand side $\phi_1, \ldots, \phi_m$ is *the antecedent* of the sequent (8.1), and the right-hand side $\psi_1, \ldots, \psi_n$ is its *succedent*. The formulas $\phi_1, \ldots, \phi_m$ are *antecedent members* of the sequent (8.1) and the formulas $\psi_1, \ldots, \psi_n$ are its *succedent members*. Antecedent and succedent are considered as multisets, that is, the order of formulas is disregarded. If Γ and Δ are multisets of formulas, then Γ, Δ is the result of concatenation, and Γ, ϕ means $\Gamma, \{\phi\}$, as before. We always treat $\leftrightarrow$ as an abbreviation: $\phi \leftrightarrow \psi := (\phi \rightarrow \psi)\&(\psi \rightarrow \phi)$.

Propositional system LJpm

Axioms:

$\phi \Rightarrow \phi$ $\qquad\qquad$ $\bot \Rightarrow p$ for atomic formulas p

Inference rules:

$$\frac{\Gamma \Rightarrow \Delta, \phi \quad \Gamma \Rightarrow \Delta, \psi}{\Gamma \Rightarrow \Delta, \phi \& \psi} \qquad \frac{\phi, \psi, \Gamma \Rightarrow \Delta}{\phi \& \psi, \Gamma \Rightarrow \Delta}$$

$$\frac{\phi, \Gamma \Rightarrow \Delta \quad \psi, \Gamma \Rightarrow \Delta}{\phi \vee \psi, \Gamma \Rightarrow \Delta} \qquad \frac{\Gamma \Rightarrow \Delta, \phi, \psi}{\Gamma \Rightarrow \Delta, \phi \vee \psi}$$

$$\frac{\phi \to \psi, \Gamma \Rightarrow \Delta, \phi \quad \psi, \Gamma \Rightarrow \Delta}{\phi \to \psi, \Gamma \Rightarrow \Delta} \qquad \frac{\phi, \Gamma \Rightarrow \psi}{\Gamma \Rightarrow \Delta, \phi \to \psi}$$

$$\frac{\phi, \phi, \Gamma \Rightarrow \Delta}{\phi, \Gamma \Rightarrow \Delta} \; contr \; \frac{\Gamma \Rightarrow \Delta, \phi, \phi}{\Gamma \Rightarrow \Delta, \phi} \qquad \frac{\Gamma \Rightarrow \Delta}{\phi, \Gamma \Rightarrow \Delta} \; weak \; \frac{\Gamma \Rightarrow \Delta}{\Gamma \Rightarrow \Delta, \phi}$$

The calculus has eight *logical rules*, namely, two rules for each connective **c**: One rule introduces it to the succedent, and it is called ($\Rightarrow$ **c**) or **c**-succedent; the second rule introduces **c** in the antecedent, and it is called (**c**$\Rightarrow$) or **c**-antecedent. Contraction *contr* and weakening *weak* are *structural rules.*

All rules except $\Rightarrow\to$ have the same *parametric* formulas Γ, Δ in conclusion and all premises.

If negation is taken as a separate connective, corresponding rules become

$$\frac{\phi, \Gamma \Rightarrow}{\Gamma \Rightarrow \Delta, \neg\phi} \qquad \frac{\neg\phi, \Gamma \Rightarrow \Delta, \phi}{\neg\phi, \Gamma \Rightarrow \Delta}$$

An additional rule that is not officially part of **LJpm** but is proved to be admissible later is needed to establish connections with other formalizations. This is a cut rule:

$$\frac{\Gamma \Rightarrow \Delta \quad \phi, \Gamma \Rightarrow \Delta}{\Gamma \Rightarrow \Delta} \; cut$$

Sequents written over the line in a rule are called *premises*; the sequent under the line is the *conclusion* of the rule. A formula explicitly shown in the conclusion of the rule and containing the connective introduced by the rule is called the *principal formula.* Thus $\phi \& \psi$ is the principal formula for &-rules, $\phi \vee \psi$ for $\vee$-rules, and so on. The subformulas ϕ, ψ of the main formula explicitly shown in the premises are *side formulas.* Formulas in the lists Γ, Δ are *parametric formulas*; the formula ϕ in the cut rule is the *cut formula.*

Contraction allows us to treat multisets in the antecedent and succedent as sets, that is, to disregard not only order of formulas, but also the number of occurrences. Weakening allows us to add arbitrary formulas to antecedent and succedent, in particular to treat $\Gamma, \phi \Rightarrow \phi, \Delta$ and $\Gamma, \bot \Rightarrow \Delta$ as axioms and derive pruned forms of one-premise rules:

$$\frac{\phi, \Gamma \Rightarrow \Delta}{\phi \& \psi, \Gamma \Rightarrow \Delta} \quad \frac{\Gamma \Rightarrow \Delta, \phi}{\Gamma \Rightarrow \Delta, \phi \vee \psi} \quad \frac{\Gamma \Rightarrow \psi}{\Gamma \Rightarrow \Delta, \phi \to \psi} \quad \frac{\phi, \Gamma \Rightarrow}{\Gamma \Rightarrow \Delta, \phi \to \psi}$$

Using contraction, we can preserve main formulas in the premises of rules or omit these formulas. For example, such inferences as:

$$\frac{\phi \& \psi, \phi, \psi, \Gamma \Rightarrow \Delta}{\phi \& \psi, \Gamma \Rightarrow \Delta}$$

are derivable:

$$\dfrac{\dfrac{\phi\&\psi, \phi, \psi, \Gamma \Rightarrow \Delta}{\phi\&\psi, \phi\&\psi, \Gamma \Rightarrow \Delta}}{\phi\&\psi, \Gamma \Rightarrow \Delta}$$

System LJpm differs from sequent formalizations of the traditional (classical) logic LK mainly in the formulation of the succedent rules for $\neg$ and $\rightarrow$, where succedent parametric formulas do not occur in the premise (although they occur in conclusion). Note that the classical forms of these rules:

$$\dfrac{\phi, \Gamma \Rightarrow \Delta}{\Gamma \Rightarrow \Delta, \neg\phi} \qquad \dfrac{\phi, \Gamma \Rightarrow \Delta, \psi}{\Gamma \Rightarrow \Delta, \phi \rightarrow \psi}$$

allow us to derive $p \vee \neg p$, so these are not admissible intuitionistically.

A *derivation* or *proof* in LJpm is defined in a standard way as a *tree* proceeding from the axioms by application of inference rules.

Example (derivations in LJpm).

$$\dfrac{\dfrac{\dfrac{\dfrac{\dfrac{q \Rightarrow q}{p, q \Rightarrow q}}{p\&q \Rightarrow q} \quad \dfrac{\dfrac{p \Rightarrow p}{p, q \Rightarrow p}}{p\&q \Rightarrow p}}{p\&q \Rightarrow q\&p}}{\Rightarrow p\&q \rightarrow q\&p} \qquad \dfrac{\dfrac{p \Rightarrow p}{p, p \rightarrow q \Rightarrow p} \quad q \Rightarrow q}{p, p \rightarrow q \Rightarrow q}$$

LEMMA 8.1. *Every rule of LJpm has a subformula property: Each formula in a premise is a subformula of some formula in the conclusion. Every derivation consists of subformulas of its last sequent.*

Proof. Check all rules. ⊣

Note that the cut rule does not have the subformula property. The systems NJp and LJpm are equivalent; moreover their rules are in exact correspondence.

LEMMA 8.2. *Let* $\Gamma \Rightarrow \Delta$ *be translated as* $\Gamma \Rightarrow \vee\Delta$. *Then a sequent is derivable in LJpm plus cut if and only if its translation is derivable in NJp. In particular* $\Gamma \vdash \phi$ *in LJpm plus cut iff* $\Gamma \vdash \phi$ *in NJp. More precisely,*

(a) every introduction rule is the succedent logical rule of LJpm for the same connective;

(b) every elimination rule can be translated by the antecedent rule for the same connective plus cut;

(c) translation of every rule of LJpm is derivable in NJp.

Proof. Consider only several subcases. Part (b) :

$\&E$:

$$\dfrac{\Gamma \Rightarrow \phi\&\psi \quad \dfrac{\dfrac{\phi \Rightarrow \phi}{\phi, \psi \Rightarrow \phi}}{\phi\&\psi \Rightarrow \phi}}{\Gamma \Rightarrow \phi}$$

$\to E$:

$$\frac{\Gamma \Rightarrow \phi \to \psi \qquad \dfrac{\dfrac{\Gamma \to \phi}{\phi \to \psi, \Gamma \Rightarrow \phi} \quad \dfrac{\psi \Rightarrow \psi}{\Gamma, \psi \Rightarrow \psi}}{\phi \to \psi, \Gamma \Rightarrow \psi}}{\Gamma \Rightarrow \psi}$$

Part (c): Consider translations of the rules of LJpm. Note that every succedent logical rule of LJpm is translated using the introduction rule for the same connective, and every antecedent logical rule of LJpm is translated using the elimination rule for the same connective combined with cut. We write translations of $\Rightarrow\to, \Rightarrow \&, \to\Rightarrow$:

$$\dfrac{\dfrac{\phi, \Gamma \Rightarrow \psi}{\Gamma \Rightarrow \phi \to \psi}}{\Gamma \Rightarrow \delta \vee (\phi \to \psi)}$$

$$\frac{\Gamma \to \delta \vee \phi \qquad \dfrac{\delta \to \delta}{\delta \to \delta \vee (\phi \& \psi)} \qquad \dfrac{\Gamma \Rightarrow \delta \vee \psi \quad \dfrac{\delta \Rightarrow \delta}{\delta \Rightarrow \delta \vee (\phi \& \psi)} \quad \dfrac{\dfrac{\phi \Rightarrow \phi \quad \psi \Rightarrow \psi}{\phi, \psi \Rightarrow \phi \& \psi}}{\phi, \psi \Rightarrow \delta \vee (\phi \& \psi)}}{\Gamma, \phi \to \delta \vee (\phi \& \psi)}}{\Gamma \to \delta \vee (\phi \& \psi)}$$

$$\frac{\Gamma \Rightarrow \phi \vee \delta \qquad \dfrac{\phi, \phi \to \psi \Rightarrow \psi \quad \dfrac{\psi, \Gamma \Rightarrow \delta}{\Gamma \Rightarrow \psi \to \delta}}{\phi, \phi \to \psi, \Gamma \Rightarrow \delta} \qquad \delta \to \delta}{\phi \to \psi \Rightarrow \delta}$$

EXERCISE 8.1. Derive following sequents in LJpm:

1. $p \vee q \Rightarrow q \vee p$
2. $p \Rightarrow q \to p$
3. $p \to q, (p \to q) \to p \Rightarrow p$
4. $p, p \to q, q \to r \Rightarrow r$
5. $(p \vee (p \to q)) \to q \Rightarrow q$
6. $(((p \to q) \to p) \to p) \to q \Rightarrow q$

EXERCISE 8.2. Construct Kripke countermodels for the following sequents:

1. $\neg\neg p \Rightarrow p$
2. $(p \to q) \to p \Rightarrow p$
3. $(p \to q) \vee (q \to p)$
4. $(\neg p \to (q \vee r)) \Rightarrow (\neg p \to q) \vee (\neg p \to r)$

8.1. Soundness of the System LJpm

The value of a sequent in a model is defined exactly as for the corresponding formula:

$V(\alpha_1, \ldots, \alpha_m \Rightarrow \beta_1, \ldots \beta_n, w) = 1$ iff for any w' such that Rww'
$V(\alpha_1, w') = \ldots V(\alpha_m, w') = 1$ implies $V(\beta_j, w') = 1$ for some $j \leq n$.

THEOREM 8.1. *Cut and all the rules of LJpm are sound: If all premises are true in a world w of a Kripke model, then the conclusion is also true in w.*

Proof. Use Theorem 7.1. and Lemma 8(c). ⊣

Note. All LJpm rules except the succedent rules for $\neg$ and $\rightarrow$ are *invertible*: Derivability of the conclusion implies derivability of the premises. This can be verified directly or inferred from the following observation: The premises are derivable from the conclusion using cut. This also shows that the truth of the conclusion implies the truth of premises.

EXERCISE 8.3. Verify that all rules except $\Rightarrow \neg$ and $\Rightarrow\rightarrow$ are invertible. Prove that $\Rightarrow \neg$ and $\Rightarrow\rightarrow$ are not invertible.

8.2. Completeness and Admissibility of Cut

DEFINITION 8.1. *A sequent $\Gamma \Rightarrow \Delta$ is* falsified *in a world w of a Kripke model (W, R, V) if $V(\&\Gamma, w) = 1, V(\vee\Delta, w) = 0$. This implies $V(\Gamma \Rightarrow \Delta, w) = 0$.*

In this Chapter $\Gamma \vdash \Delta$ means that $\Gamma \Rightarrow \Delta$ is derivable in LJpm.

We now prove that any sequent underivable in LJpm is falsified in some world (of some model). Moreover we prove (following [6]) that there is a universal (canonical) model suitable for the falsification of all underivable sequents (each in its own world). It is essential that this model proves completeness of a cut-free formulation.

This proof is a natural extension of a completeness proof for classical predicate logic.

DEFINITION 8.2. *$Sub(\Gamma)$ stands for the set of all subformulas of Γ. A sequent $\Gamma \Rightarrow \Delta$ is* complete *if it is underivable and for any formula $\phi \in Sub(\Gamma, \Delta)$ either $\phi \in \Gamma \cup \Delta$ or*

$$\Gamma \vdash \Delta, \phi \text{ and } \phi, \Gamma \vdash \Delta, \tag{8.3}$$

that is, both sequents $\Gamma \Rightarrow \Delta, \phi$ and $\phi, \Gamma \Rightarrow \Delta$ are derivable.

Note. The condition (8.3) is used in the proof of the Corollary 8.2., but that Corollary shows that (8.3) never holds if $\Gamma \Rightarrow \Delta$ is underivable.

The following definitions are motivated in part by the proof-search process, which proceeds by bottom-up application of the inference rules.

DEFINITION 8.3. *A sequent $\Gamma \Rightarrow \Delta$ is* saturated for invertible rules *if it is underivable and the following conditions are satisfied for any ϕ, ψ:*

$(\Rightarrow \&)$ $(\phi \& \psi) \in \Delta$ *implies* $\phi \in \Delta$ *or* $\psi \in \Delta$.

$(\& \Rightarrow)$ $(\phi \& \psi) \in \Gamma$ *implies* $\phi, \psi \in \Gamma$.

$(\Rightarrow \vee)$ $(\phi \vee \psi) \in \Delta$ *implies* $\phi, \psi \in \Delta$.

$(\vee \Rightarrow)$ $(\phi \vee \psi) \in \Gamma$ *implies* $\phi \in \Gamma$ *or* $\psi \in \Gamma$.

$(\rightarrow\Rightarrow)$ $(\phi \rightarrow \psi) \in \Gamma$ *implies* $\phi \in \Delta$ *or* $\psi \in \Gamma$.

The following Lemma says that all invertible rules were applied from the bottom up in every complete sequent.

LEMMA 8.3. *(saturation) If* $\Gamma \Rightarrow \Delta$ *is complete, then it is saturated for invertible rules.*

Proof. We leave all but three cases to the reader. We say that ϕ *clashes* with $\Gamma \Rightarrow \Delta$ if (8.3) holds; that is, both $\Gamma \Rightarrow \Delta, \phi$ and $\phi, \Gamma \Rightarrow \Delta$ are derivable.

$(\Rightarrow \&)$: Let $\phi \& \psi \in \Delta$. We analyze all cases for three possible positions of either ϕ, ψ in Γ, in Δ, or in clash with $\Gamma \Rightarrow \Delta$. Note that if

$$\Gamma \vdash \Delta, \phi \text{ and } \Gamma \vdash \Delta, \psi \tag{8.4}$$

then $\Gamma \Rightarrow \Delta$ is derivable by $\Rightarrow \&$, which contradicts completeness of $\Gamma \Rightarrow \Delta$.

Hence one of ϕ, ψ (say ϕ) is not in Γ [otherwise the sequents (8.4) are axioms], and one of ϕ, ψ does not clash with $\Gamma \Rightarrow \Delta$. If $\phi \notin \Delta$, then ϕ clashes with $\Gamma \Rightarrow \Delta$ by the completeness of $\Gamma \Rightarrow \Delta$. Hence the sequent $\Gamma \Rightarrow \Delta, \phi$ is derivable and ψ does not clash with $\Gamma \Rightarrow \Delta$. Also $\psi \notin \Gamma$, since otherwise sequents in (8.4) are derivable; hence $\psi \in \Delta$ as required.

$(\& \Rightarrow)$: Let $\phi \& \psi \in \Gamma$. Then $\phi \notin \Delta$, since otherwise $\Gamma \Rightarrow \Delta$ is derivable by $(\& \Rightarrow)$. If ϕ clashes with $\Gamma \Rightarrow \Delta$, then $\Gamma \Rightarrow \Delta$ is derivable from $\phi, \Gamma \Rightarrow \Delta$ by weakening, $(\& \Rightarrow)$, and contraction. For ϕ this leaves only the possibility that $\phi \in \Gamma$. For analogous reasons, $\psi \in \Gamma$.

$(\rightarrow\Rightarrow)$: Let $(\phi \rightarrow \psi) \in \Gamma$. As before one of the following sequents is underivable:

$$\Gamma \Rightarrow \Delta, \phi, \quad \psi, \Gamma \Rightarrow \Delta, \tag{8.5}$$

since $\Gamma \Rightarrow \Delta$ is derivable from (8.5) by $\rightarrow\Rightarrow$. Hence one of ϕ, ψ does not clash with $\Gamma \Rightarrow \Delta$, and ($\phi \notin \Gamma$ *or* $\psi \notin \Delta$).

Assume that $\phi \notin \Gamma$. If $\phi \notin \Delta$, then ϕ clashes with $\Gamma \Rightarrow \Delta$ by completeness, and ψ does not clash with $\Gamma \Rightarrow \Delta$. Hence $\psi \in \Gamma$, as required.

In the remaining case ($\phi \in \Gamma$ *and* $\psi \notin \Delta$), if $\psi \notin \Gamma$, then ψ clashes with $\Gamma \Rightarrow \Delta$, so ϕ does not clash with $\Gamma \Rightarrow \Delta$. Hence $\phi \in \Delta$, as required. ⊣

EXERCISE 8.4. Verify all remaining cases.

LEMMA 8.4. *(completion). Any underivable sequent* $\Gamma_0 \Rightarrow \Delta_0$ *can be extended to a complete sequent consisting of subformulas of* Γ_0, Δ_0.

Proof. Consider an enumeration $\phi_0, \phi_1, \ldots$ of all (propositional) formulas in $Sub(\Gamma_0, \Delta_0)$. Define the sequences $\Gamma_0 \subset \Gamma_1 \subset \ldots, \Delta_0 \subset \Delta_1 \subset \ldots$ of finite sets of formulas such that $\Gamma_i \Rightarrow \Delta_i$ is underivable and complete for all formulas ϕ_j, $j < i$: Either $\phi_j \in \Gamma_i \cup \Delta_i$ or both $\Gamma_i \Rightarrow \Delta_i, \phi_j$ and $\phi_j, \Gamma_i \Rightarrow \Delta_i$ are derivable.

Let $\Gamma_{i+1} := \Gamma_i \cup \{\phi_i\}$ if $\phi_i, \Gamma_i \Rightarrow \Delta_i$ is underivable; otherwise let $\Gamma_{i+1} = \Gamma_i$. Then let $\Delta_{i+1} := \Delta_i \cup \{\phi_i\}$ if $\Gamma_{i+1} \Rightarrow \Delta_i, \phi_i$ is underivable; otherwise let $\Delta_{i+1} = \Delta_i$. Let $\Gamma =: \bigcup \Gamma_i$ and $\Delta := \bigcup \Delta_i$. The completeness of $\Gamma \Rightarrow \Delta$ easily follows from the completeness of $\Gamma_i \Rightarrow \Delta_i$ for $\{\phi_0, \ldots, \phi_i\}$. ⊣

DEFINITION 8.4. *Consider the following Kripke model:* $\mathbf{K} = \langle W, R_{\subseteq}, V_{\in} \rangle$:

- *W is the set of all complete sequents.*
- $R_{\subseteq}(\Gamma \Rightarrow \Delta, \Gamma' \Rightarrow \Delta')$ *iff* $\Gamma \subseteq \Gamma'$.
- $V_{\in}(p, \Gamma \Rightarrow \Delta) = 1$ *iff* $p \in \Gamma$.

$\mathbf{K}$ is clearly a Kripke model: $R_{\subseteq}$ is reflexive and transitive, since $\subseteq$ is reflexive and transitive; $V_{\in}$ is monotonic, since $\in$ is monotonic with respect to $\subseteq$.

We prove below that $\mathbf{K}$ falsifies every invalid formula.

Let us state the properties of $\mathbf{K}$, which we will need later.

DEFINITION 8.5. *A set M of sequents is* saturated for non-invertible rules *if the following condition is satisfied for every $\Gamma \Rightarrow \Delta$ in M:*

($\Rightarrow\rightarrow$) *if $\phi \rightarrow \psi \in \Delta$, then there is a sequent $\Gamma' \Rightarrow \Delta'$ in M such that $\phi, \Gamma \subseteq \Gamma'$ and $\psi \in \Delta'$.*

A set M of sequents is saturated *if every sequent in M is saturated for invertible rules and M is saturated for non-invertible rules.*

Comment. In a saturated set of sequents, both invertible and non-invertible rules are eventually applied from the bottom up.

LEMMA 8.5. *The set W of all complete sequents is saturated*

Proof. Only condition $\Rightarrow\rightarrow$ needs to be checked. If $\Gamma \Rightarrow \Delta, \phi \rightarrow \psi$ is in W, then the sequent $\phi, \Gamma \Rightarrow, \psi$ is underivable. By the Completion Lemma 8.2., we can be extend this sequent to a complete sequent $w' = \Gamma' \Rightarrow \Delta'$; hence $\phi, \Gamma \subseteq \Gamma'$, and $\psi \in \Delta'$ as required. ⊣

DEFINITION 8.6. A model defined by a saturated set M *is* $K_M = \langle M, R_M, V_M \rangle$, *where* $R_M = R_{\subseteq}$, $V_M = V_{\in}$ *as defined above.*

THEOREM 8.2. *Let M be a saturated set. Then for $w \equiv \Gamma \Rightarrow \Delta$, $w \in M$:*

$$\theta \in \Gamma \text{ implies } V_M(\theta, w) = 1 \tag{8.6}$$

$$\theta \in \Delta \text{ implies } V_M(\theta, w) = 0 \tag{8.7}$$

$$V_M(\Gamma \Rightarrow \Delta, w) = 0, \quad \text{that is, } \Gamma \Rightarrow \Delta \text{ is falsified in the world } w \text{ of } K_M \tag{8.8}$$

Proof. Relation (8.8) is an immediate consequence of (8.6,8.7), which are proved by simultaneous induction on the formula θ.

$\theta = p$ (induction base):

In this case (8.6) follows from the definition of V_M. If $\theta \in \Delta$, then $\theta \notin \Gamma$; otherwise $\Gamma \Rightarrow \Delta$ is an axiom of LJpm. Hence $V_M(\theta, w) = 0$.

The induction step (θ is a composite formula) is proved by cases.

&: $\theta = \phi \& \psi$. If $\theta \in \Gamma$, then $\phi, \psi \in \Gamma$ by the saturation for invertible rules; therefore $V_M(\phi, w) = V_M(\psi, w) = 1$ by the induction hypothesis. Hence $V_M(\theta, w) = 1$ by the truth condition for &.

If $\theta \in \Delta$, then one of ϕ, ψ, say, ϕ is in Δ, by the saturation condition. Hence $V_M(\phi, w) = 0$ by the induction hypothesis; therefore $V_M(\phi \& \psi, w) = 0$.

$\rightarrow$: $\theta = \phi \rightarrow \psi$. If $\theta \in \Gamma$, then for every $w' \equiv \Gamma' \Rightarrow \Delta'$ such that $R_M\, w\, w'$ (that is, $\Gamma \subseteq \Gamma'$) we have $\theta \in \Gamma'$. By the saturation condition, this implies $\phi \in \Delta'$ or $\psi \in \Gamma'$. By the induction hypothesis, $V_M(\phi, w') = 0$ or $V_M(\psi, w') = 1$. This implies $V_M(\phi \rightarrow \psi, w) = 1$.

If $\theta \in \Delta$, then by the saturation condition ($\Rightarrow\rightarrow$) we have $w' \equiv \Gamma' \Rightarrow \Delta'$ in M, such that $R_M\, w\, w'$, $\phi \in \Gamma'$, $\psi \in \Delta'$. By the induction hypothesis, $V(\phi, w') = 1$ and $V(\psi, w') = 0$. This implies $V(\phi \rightarrow \psi, w) = 0$, as required.

The remaining cases are simpler. ⊣

EXERCISE 8.5. Verify the remaining cases.

COROLLARY 8.1. *(completeness) (a) Each sequent underivable in LJpm is falsified in the canonical model* **K**. *Hence every valid sequent is derivable in LJpm.*

(b) Moreover every underivable sequent $\Gamma \Rightarrow \Delta$ is falsified in a finite model $\mathbf{K}_M$, *where $M = M_{\Gamma \Rightarrow \Delta}$ is the set of all complete sequents consisting of subformulas of Γ, Δ. The number of elements $\|M_{\Gamma \Rightarrow \Delta}\|$ is bounded by 4^s, where* $s = \|Sub(\Gamma, \Delta)\|$.

Proof. Part (a): By the Completion Lemma 8.2., any underivable sequent $\Gamma \Rightarrow \Delta$ can be extended to a complete sequent $w = \Gamma' \Rightarrow \Delta'$; by the Theorem 8.2., $V(\Gamma' \Rightarrow \Delta', w) = 0$. Hence $V(\Gamma \Rightarrow \Delta, w) = 0$ by monotonicity.

Part (b): All arguments in the proof of Part (a) remain valid for $\mathbf{K}_M$ instead of **K**. Let us estimate $\|M_{\Gamma \Rightarrow \Delta}\|$. For every $\Sigma \Rightarrow \Pi \in M_{\Gamma \Rightarrow \Delta}$, we have $\|\Sigma\| \leq s$, $\|\Pi\| \leq s$; hence the number of possibilities for Σ and for Π is bounded by 2^s, so $\|M_{\Gamma \Rightarrow \Delta}\| \leq 2^s 2^s = 4^s$. ⊣

THEOREM 8.3. *(soundness and completeness).*

(a) A formula is derivable in LJpm *iff it is valid.*

(b) A formula is derivable in LJpm *if and only if it is valid in all finite pointed models where accessibility relation is a partial order.*

Proof. Part (a): Apply Theorem 8.1. and Corollary 8.2.(a).

Part (b): Apply Corollary 8.2.(b) and note that the construction in the proof of Lemma 8.2. preserves finiteness. ⊣

COROLLARY 8.2. *(a) Cut rule is admissible in LJpm: If* $\Gamma \Rightarrow \Delta, \phi$ *and* $\phi, \Gamma \Rightarrow \Delta$ *are derivable, then* $\Gamma \Rightarrow \Delta$ *is derivable.*

(b) LJpm is equivalent to NJp with respect to derivability of sequents $\Gamma \Rightarrow \alpha$

Proof. Part (a): By the soundness of LJpm plus cut, it follows that $\Gamma \Rightarrow \Delta$ is valid. By the completeness of LJpm (without cut) it is derivable.

Part (b): Use Lemma 8 ⊣

Comment. What cut elimination method is suggested by this Corollary? Given (cutfree) derivations of the premises, ignore them to search for a cutfree proof of $\Gamma \Rightarrow \Delta$. You will eventually find it; as noted in the introduction, there are more sophisticated methods.

8.3. Translation into the Predicate Logic

The completeness theorem 8.2. for LJpm allows us to write a predicate formula that expresses the validity of a given propositional formula φ. To every propositional variable p, we assign a monadic predicate variable P. We also fix a binary predicate symbol R.

DEFINITION 8.7. *For any propositional formula* φ *and individual variable* w*, we define a predicate formula* (φ, w) *by induction on* φ*:*

$$\begin{array}{lll} (p, w) & := & P(w) \\ (\varphi \& \psi, w) & := & (\varphi, w) \& (\psi, w) \\ (\varphi \vee \psi, w) & := & (\varphi, w) \vee (\psi, w) \\ (\varphi \to \psi, w) & := & \forall w'(R(w, w') \& (\varphi, w') \to (\psi, w')) \end{array} \tag{8.9}$$

where w' *is a new variable.*

DEFINITION 8.8.

$$\varphi^P := \kappa \to (\phi, a)$$

where:

$$\kappa := (\forall w R(w, w) \& \forall w \forall w_1 \forall w_2 (R(w, w_1) \& R(w_1, w_2) \to R(w, w_2)) \&$$

$$\&_i \forall w \forall w_1 (R(w, w_1) \to (P_i(w) \to P_i(w_1)))$$

where w, w_1, w_2, a *are distinct new individual variables and* p_i *are all predicate variables occurring in* ϕ*.*

THEOREM 8.4. *A propositional formula* φ *is derivale in LJpm iff* φ^P *is derivable in the classical predicate calculus.*

Proof.A first-order model:

$$M = \langle W, R, P_1, \ldots, P_n, a_0 \rangle$$

with the universe W and an interpretation for the predicates and the constant in φ^P, satisfying the premise of φ^P (that is, reflexivity and transitivity of R) generates a Kripke model:

$$M' = \langle W, R, V_{M'} \rangle$$

where $V_{M'}(p_i, w) = 1$ iff $M \models P_i(w)$ for $w \in W$.

In the other direction, any Kripke model M' with a distinguished world a_0 generates a first-order model M by means of (8.9); that is, $M \models P(w)$ iff $M' \models (p, w) = 1$. Moreover by induction on ψ, it is easy to prove that:

$$M \models \psi^P[w] \quad \text{iff} \quad V_{M'}(\psi, w) = 1 \tag{8.10}$$

We see that φ^P is valid (in the first-order logic) iff φ is valid intuitionistically. $\dashv$

This may initially suggest an alternative proof-search procedure. To test φ, write φ^P and apply some (resolution) theorem prover for the classical logic. However the result will be devastating. The search space is very quickly filled by useless resolvents. Much more sophisticated approach along these lines is proposed in [18].

8.4. Algebraic Models

Algebraic models represent another kind of model for non-classical logics. Such a model usually consists of a universe U with a distinguished element $\top$ (truth) and operations on U corresponding to logical operations. Conditions for operations on U are set so that the validity of derivable formulas can be proved rather easily. We show here that intuitionistic propositional logic is sound and complete for algebraic models.

DEFINITION 8.9. *A pseudo-Boolean algebra (PBA) or Heyting algebra is a pair $(\mathcal{B}, \leq)$ where $\mathcal{B}$ is a non-empty set and $\leq$ is a partial ordering relation on $\mathcal{B}$, which defines a lattice with the least element $\perp$ and psudo-complements.*

This means that for arbitrary a and b:

(i) the least upper bound $a \sqcup b$,

(ii) the greatest lower bound $a \sqcap b$, and

(iii) the pseudo-complement $a \to b := \max\{c \mid c \sqcap a \leq b\}$ always exist. Lattice conditions imply in addition to (i), (ii) commutativity and associativity of $\sqcup, \sqcap$, and distributivity.

To make this definition easier to remember, associate $a \leq b$ with the truth of $a \to b$, $\sqcap$ with & and $\sqcup$ with $\vee$. Then the conditions for pseudo-complements say that for arbitrary a, b, and c:

$$(a \sqcap (a \to b)) \to b = \top, \qquad ((c \sqcap a) \to b) \to (c \to (a \to b)) = \top$$

The first equation corresponds to modus ponens, establishing that $a \sqcap (a \rightarrow b) \leq b$; that is, it enforces suitable constraints on $a \rightarrow b$. The second equation establishes that if $a \sqcap c \leq b$, then $c \leq (a \rightarrow b)$; that is, the maximality of $a \rightarrow b$.

Truth $\top$ is defined as $\bot \rightarrow \bot$, negation as $-a = a \rightarrow \bot$.

A *valuation* from the set $\mathcal{P}$ of propositional variables to a PBA $\langle \mathcal{B}, \leq \rangle$ is a function $h : \mathcal{P} \rightarrow \mathcal{B}$. The valuation h is extended to all formulas as a homomorphism:

$$\begin{array}{lcl} h(\alpha\&\beta) & = & h(\alpha) \sqcap h(\beta) \\ h(\alpha \vee \beta) & = & h(\alpha) \sqcup h(\beta) \\ h(\alpha \rightarrow \beta) & = & h(\alpha) \rightarrow h(\beta) \\ h(\bot) & = & \bot \end{array}$$

Formula α is *true in a PBA* $\langle \mathcal{B}, \leq \rangle$ and a valuation h iff $h(\alpha) = \top$. Then we say that h satisfies α. Formula α is *valid in a PBA* if it is true in any valuation to this PBA. The α is *valid* if it is valid in any PBA. The sequent $\Gamma \Rightarrow \Delta$ is valid iff the formula $\&\Gamma \rightarrow \vee\Delta$ is valid, that is, $\sqcap_{\gamma \in \Gamma} h(\gamma) \leq \sqcup_{\delta \in \Delta} h(\delta)$, or for short $h(\&\Gamma) \leq h(\vee\Delta)$.

LEMMA 8.6. *LJpm is sound: every derivable sequent is valid in any PBA.*

Proof. Let us verify that axioms are valid and inference rules preserve truth in every PBA and every evaluation h.

Axioms: $\alpha, \Gamma \Rightarrow \Delta, \alpha \quad h(\alpha) \sqcap h(\&\Gamma) \leq h(\alpha) \leq h(\vee\Delta) \sqcup h(\alpha)$

Inference rules $(\& \Rightarrow, \Rightarrow \&), (\vee \Rightarrow, \Rightarrow \vee)$ reflect corresponding properties of $\sqcup$, $\sqcap$; for example,

$(\Rightarrow \&)$:

$$\frac{\Gamma \Rightarrow \Delta, \varphi \qquad \Gamma \Rightarrow \Delta, \psi}{\Gamma \Rightarrow \Delta, \varphi\&\psi}.$$

Let $h(\&\Gamma) \leq h(\vee\Delta) \sqcup h(\varphi)$, $h(\&\Gamma) \leq h(\vee\Delta) \sqcup H(\psi)$. Then:

$$h(\&\Gamma) \leq (h(\vee\Delta) \sqcup h(\varphi)) \sqcap (h(\vee\Delta) \sqcup h(\psi)) \leq h(\vee\Delta) \sqcup h(\varphi \sqcap \psi)$$

by distributivity.

$(\Rightarrow\rightarrow)$:

$$\frac{\Gamma, \varphi \Rightarrow \psi}{\Gamma \Rightarrow \Delta, \varphi \rightarrow \psi}$$

If $h(\varphi) \sqcap h(\&\Gamma) \leq h(\psi)$, then $h(\&\Gamma) \leq h(\varphi) \rightarrow h(\psi) \leq h(\vee\Delta) \sqcup (h(\varphi) \rightarrow h(\psi))$

$(\rightarrow\Rightarrow)$:

$$\frac{\varphi \rightarrow \psi, \Gamma \Rightarrow \Delta, \varphi \qquad \psi, \Gamma \Rightarrow \Delta}{\varphi \rightarrow \psi, \Gamma \Rightarrow \Delta}$$

Let $h(\varphi \rightarrow \psi) \sqcap h(\&\Gamma) \leq h(\vee\Delta) \sqcup h(\varphi)$, $h(\psi) \sqcap h(\&\Gamma) \leq h(\vee\Delta)$. Then:

$$h(\varphi \rightarrow \psi) \sqcap h(\&\Gamma) \leq h(\vee\Delta) \sqcup (h(\varphi) \sqcap h(\varphi \rightarrow \psi)) \leq h(\vee\Delta) \sqcup h(\psi)$$

$$h(\varphi \rightarrow \psi) \sqcup h(\&\Gamma) \leq h(\vee\Delta) \sqcup (h(\psi) \sqcap h(\&\Gamma)) \leq h(\vee\Delta) \sqcap h(\vee\Delta) = h(\vee\Delta).$$

⊣

To prove completeness of the algebraic semantics, we translate Kripke models into algebraic models.

DEFINITION 8.10. *Let $K = \langle W, R\rangle$ be Kripke frame, that is, a non-empty set W plus a reflexive, transitive relation R. A set $b \subseteq W$ is R-closed iff $w \in b$ and Rww' imply $w' \in b$.*

We define $\mathcal{B} = \mathcal{B}_K$ to be the collection of R-closed subsets of W with the inclusion relation $\subseteq$ as the ordering $\leq$.

NOTE. The R-closed sets are possible values of propositional variables in K for elements of $\mathcal{B}$. Union and intersection of R-closed sets are R-closed. The set of all worlds, $\top = W$, is R-closed as well as $\emptyset = \bot$.
The main step in proving that $\langle \mathcal{B}, \leq\rangle$ is a PBA is to define pseudo-complements. We define $M \to N$ to be the largest R-closed subset of $(W \setminus M) \cup N$.

LEMMA 8.7. *(Properties of operations).*

(a) $\cup, \cap$ are lattice join and meet (lattice union and intersection);

(b) $\to$ is the relative pseudo-complement with respect to $\cap, \cup$;

(c) $\emptyset$ is the least element.

Proof. Parts (a) and (c) are easy; let us prove (b).

1. The first half of the pseudo-complement condition $M \cap (M \to N) \subseteq N$ is obtained as follows:

$$\begin{aligned} M \cap (M \to N) &\subseteq M \cap [(W \setminus M) \cup N] \\ &= [M \cap (W \setminus M)] \cup [M \cap N] \\ &= M \cap N \\ &\subseteq N \end{aligned}$$

2. Let us prove that $M \to N$ is the greatest element X of $\mathcal{B}$ satisfying $M \cap X \subseteq N$. Indeed

if $M \cap X \subseteq N$ and $X \in \mathcal{B}$, then by monotonicity of $\cup$:
$(M \cap X) \cup (X \setminus M) \subseteq N \cup (X \setminus M)$. Since $X = (M \cap X) \cup (X \setminus M)$
$X \subseteq N \cup (X \setminus M)$, so
$X \subseteq N \cup (W \setminus M)$.

Since $X \in \mathcal{B}$, X is R-closed, hence by the definition of $M \to N$, we have $X \subseteq M \to N$; that is, $X \leq M \to N$. ⊣

LEMMA 8.8. *(correspondence between a Kripke model K and $\mathcal{B}_K$).*

Let $K = \langle W, R\rangle$ and $\langle K, V\rangle$ be a Kripke model. Then the following valuation of the propositional variables in the pseudo-Boolean algebra $\mathcal{B}_K$:

$$h(p) = \{w \mid V(p, w) = 1\} \tag{8.11}$$

extends to all formulas α*:*

$$h(\alpha) = \{w \mid V(\alpha, w) = 1\}$$

In particular α *is true in* h *iff it is true in* V *in all worlds:*

$$h(\alpha) = \top \quad \textit{iff} \quad \forall w(V(\alpha, w) = 1)$$

COROLLARY 8.3. *A formula is derivable iff it is valid in all PBA iff it is valid in all finite PBA.*

Proof. We establish Lemma 8.4. by induction on the formula α. The induction base is (8.11). The induction steps for $\&, \vee$ are obvious. The induction step for $\to$ proceeds as follows: $h(\alpha \to \beta)$ is the largest R-closed subset of $(W \setminus h(\alpha)) \cup h(\beta)$, that is, of the set:

$$\{w \mid V(\alpha, w) = 0 \text{ or } V(\beta, w) = 1\} = \{w \mid V(\alpha, w) = 1 \to V(\beta, w) = 1\}.$$

Part (a): $w \in h(\alpha \to \beta)$ implies $V(\alpha \to \beta, w) = 1$. Let $w \in h(\alpha \to \beta)$, Rww', $V(\alpha, w') = 1$. Since $h(\alpha \to \beta)$ is R-closed, $w' \in h(\alpha \to \beta)$. Then $w' \in h(\alpha)$ implies $w' \in h(\beta)$. By the induction hypothesis, this means $V(\alpha, w') = 1$ implies $V(\beta, w') = 1$.

Part (b): $V(\alpha \to \beta, w) = 1$ implies $w \in h(\alpha \to \beta)$. The set $\{w \mid V(\alpha \to \beta, w) = 1\}$ is R-closed by the monotonicity Lemma 7, and it is the largest R-closed subset of $(W \setminus \{w \mid V(\alpha, w) = 1\}) \cup \{w \mid V(\beta, w) = 1\}$. By the induction hypothesis, this is the same as the largest R-closed subset of $(W \setminus h(\alpha)) \cup h(\beta)$, equaling $h(\alpha \to \beta)$ as required. ⊣

NOTE. In fact there is an "inverse" operation constructing Kripke models from finite pseudo-Boolean algebras. For an algebra $\langle \mathcal{B}, \leq \rangle$, let W be the set of proper prime filters in $\mathcal{B}$ and let R be the inclusion relation. (A filter F satisfies the conditions $a \in F \to a \cup b \in F$ for all b; $a, b \in F \to a \cap b \in F$; a prime filter satisfies $a \cup b \in F \to a \in F$ or $b \in F$.)

The valuation V in $\langle W, R \rangle$ corresponding to a valuation h in $\mathcal{B}$ is defined by:

$$V(p, F) = 1 \quad \text{iff} \quad h(p) \in F \tag{8.12}$$

The proof that $\langle W, R \rangle$ is a Kripke model and (8.12) extends to all formulas is given in [6].

8.5. Filtration, Finite Matrices

Let us sketch another proof of Theorem 8.2.(b).

8.5.1. Filtration

For any Kripke model $M =< W, R, V >$ and a formula φ, define the *filtration* $M' =< W', R', V' >$ *of* M *through* φ as follows. Let Φ be the set of all subformulas of φ. For $w \in W$ set

$$\hat{w} = \{v \in W \mid M, w \models \psi \text{ iff } M, v \models \psi \text{ for all } \psi \in \Phi\}.$$

Set

$$W' = \{\hat{w} | w \in W\}$$

$$R'\hat{v}\hat{w} \quad \text{iff} \quad M, v \models \psi \text{ implies } M, w \models \psi \text{ for all } \psi \in \Phi$$

$$V'(p, \hat{w}) = 1 \text{ iff } V(p, w) = 1$$

Note that W' is finite, R' is reflexive and transitive, V' is monotonic, and

$$Rww' \text{ implies } R'\hat{w}\hat{w'} \tag{8.13}$$

THEOREM 8.5. *For all $\alpha \in \Phi, w \in W$:*

$$V'(\alpha, \hat{w}) = V(\alpha, w)$$

Proof. We use induction on α. Induction base and $\&, \vee$-cases are easy. Let $\alpha \equiv (\beta \to \gamma)$. Using IH and (8.13) we have:

$$V'(\alpha, \hat{w}) = 1 \text{ iff } \forall w'(R\hat{w}\hat{w'} \to (V(\beta, w') \to V(\gamma, w')))$$

$$\text{implies } \forall w'(Rww' \to (V(\beta, w') \to V(\gamma, w'))) \text{ iff } V(\alpha, w) = 1$$

Using IH and the definition of R' we have:

$$V(\alpha, w) = 1,\ R'\hat{w}\hat{w'},\ V(\beta, \hat{w'}) = 1$$

imply

$$V(\alpha, w') = 1,\ V(\beta, w') = 1,\ V(\gamma, w') = 1.$$

Therefore, $V(\alpha, w) = 1$ implies $V'(\alpha, \hat{w}) = 1$. ⊣

EXERCISE 8.6. Infer Theorem 8.2.(b) from the Theorem 8.5.1and completeness.

8.5.2. Lindenbaum Algebra

Let us outline another way of proving completeness of pseudo-Boolean algebras. This was in fact the first completeness proof [10]. Under this approach the elements of the pseudo-Boolean algebras are equivalence classes of formulas.

Define

$$\hat{\varphi} = \{\psi \mid (\varphi \leftrightarrow \psi) \text{ is derivable }\}$$

$$\mathcal{B} = \text{the set of all } \hat{\varphi}$$

$$\hat{\varphi}\&\hat{\psi} = (\widehat{\varphi\&\psi}); \qquad \hat{\varphi} \vee \hat{\psi} = (\widehat{\varphi \vee \psi})$$

EXERCISE 8.7. Verify that $(\widehat{p \to p}), (\widehat{p\&\neg p})$ are $\top, \bot$ and that $(\widehat{\varphi \to \psi})$ is a pseudo-complement. Infer the completeness theorem. The PBA $\mathcal{B}$ is called the *Lindenbaum algebra* of intuitionistic logic.

8.5.3. Finite Truth Tables

Let us now restate the definition of Kripke models in terms of *finite truth tables*. For a finite pointed Kripke frame:

$$< G, W, R > \quad \text{with } W = \{1, 2, \ldots, n\},\ G = 1$$

let every binary number $N < 2^{n+1}$ encode by its binary digits a truth value assignment to a propositional variable in W. For example, $2^{n+1} - 1 =_{binary} || \ldots |$ encodes an assignment of 1 (*true*) to our variable in all worlds, while $2^n =_{binary} |0 \ldots 0$ encodes an assignment of 1 (*true*) in the world n and 0 (*false*) in all other worlds.

Say that a number $k < 2^n$ is *monotonic* if for the ith bit $(k)_i$ of k:

$$\forall i \forall j (Rij \rightarrow (k)_i \leq (k)_j).$$

Now take the monotonic numbers among $0, 1, \ldots, 2^{n+1} - 1$ to be the truth values. Take $\&, \vee$ to be bit operations $\&, \vee$ (that is, min, max) so that for $n = 3, a = 4 = |00_{binary}, b = 6 = ||0_{binary}$:

$$a \& b = |00 \& ||0 = |00 = 4,\ a \vee b = ||0 = 6.$$

The ith bit of $a \rightarrow b$ is defined as follows to satisfy Kripke implication semantics:

$$(a \rightarrow b)_i := 1 \text{ iff } (\forall j \leq n)(Rij \rightarrow (a)_j \leq (b)_j)$$

The set of monotonic numbers among $\{0, 1, \ldots 2^{n+1} - 1\}$ with operations $\&, \vee, \rightarrow$ just defined and the distinguished value $2^{n+1} - 1$ is an example of a *finite matrix for intuitionistic logic*: A finite set of truth values with propositional operations on them given by truth tables (matrices) such that all derivable formulas are *valid*; that is, these formulas take the distinguished value under every assignment of the truth values to variables.

EXAMPLE 8.1. The Kripke model with $W = \{0, 1\}$, $R00, R11, R01$ generates a finite matrix with the truth values $\{0, 2, 3\}$: the number 1 is not monotonic. We obtain the following tables for the connectives: $\perp := 0$

&	0	2	3
0	0	0	0
2	0	2	2
3	0	2	3

∨	0	2	3
0	0	2	3
2	2	2	3
3	3	3	3

→	0	2	3
0	3	3	3
2	0	3	3
3	0	2	3

When the variable p is assigned value $2 = |0_{binary}$ corresponding to truth in the second world and the falsity in the first world, we have:

$$\neg p = (2 \rightarrow 0) = 0, p \vee \neg p = 2 \vee 0 = 2 \neq 3.$$

This means $p \vee \neg p$ is not valid.

EXERCISE 8.8. Construct a finite truth table verifying all derivable formulas but falsifying $\neg p \vee \neg\neg p$

THEOREM 8.6. *A formula is derivable in NJp iff it is valid in all finite matrices for intuitionistic logic.*

Proof. If $\not\vdash \alpha$, then take a finite pointed Kripke model falsifying α and turn it into a finite matrix. $\dashv$

Chapter 9

Topological Completeness

Let us describe topological semantics for NJp and prove a topological completeness theorem.

DEFINITION 9.1. *An* interpretation in a topological space X, *is an assignment of an* open *subset* $V(p) \subset X$ *to every propositional variable* p. *This assignment is extended to composite formulas as follows:*

$$V(\alpha \& \beta) := V(\alpha) \cap V(\beta); \qquad V(\alpha \vee \beta) := V(\alpha) \cup V(\beta); \qquad V(\bot) := \emptyset \tag{9.1}$$

$$V(\alpha \to \beta) := Interior((X - V(\alpha)) \cup V(\beta)) \tag{9.2}$$

Hence $V(\alpha)$ is open for every formula α. As a consequence:

$$V(\neg\alpha) := Interior(X - V(\alpha))$$

A formula α is *true* in the interpretation V iff $V(\alpha) = X$. A formula is *valid* in a space X if it is true in any interpretation in X.

THEOREM 9.1.

A formula is derivable in NJp iff it is valid in every topological space X.

Proof. It is easy to verify that all axioms of NJp are valid and the inference rules preserve validity. Hence every derivable formula is valid.

To prove the converse implication, assume that a formula α is not derivable in NJp and take a finite pointed Kripke model $(G, \mathbf{K}, R, \models)$ satisfying:

$$G \not\models \alpha \tag{9.3}$$

[see Corollary 8.2.(b)]. Introduce a topology on the set $\mathbf{K}$ by taking the family of all cones:

$$\{w \in \mathbf{K} \mid Rw_0w\}$$

for all $w_0 \in \mathbf{K}$ as a basis of open sets. In other words, the interior operation satisfies the following:

$$Interior(Y) := \{w \in \mathbf{K} \mid \forall w'(Rww' \to w' \in Y)\} \qquad \text{for } Y \subseteq \mathbf{K}$$

Set

$$V(p) = \{w \in \mathbf{K} \mid w \models p\}$$

The $V(p)$ is open by the monotonicity condition for $\models$, and it is easy to verify by induction on the formula γ that:

$$V(\gamma) = \{w \in \mathbf{K} \mid w \models \gamma\}.$$

Consider the most important case, $\gamma \equiv \alpha \to \beta$, and let w, w' range over $\mathbf{K}$. Then:

$$V(\alpha \to \beta) \equiv Interior((\mathbf{K} - V(\alpha)) \cup V(\beta))$$

$$\equiv Interior((\mathbf{K} - \{w \mid w \models \alpha\}) \cup \{w \mid w \models \beta\})$$

$$\equiv Interior(\mathbf{K} - \{w \mid w \models \alpha \text{ and } w \not\models \beta\})$$

$$= \{w \mid \forall w'(Rww' \to (w' \models \alpha \to w' \models \beta))\} = V(\alpha \to \beta)$$

Hence (9.3) shows that $V(\alpha) \neq X$; that is, α is not valid. ⊣

The construction used in the preceding proof yields a teratological space: For example, the closure of a singleton set $\{w\}$ is the set of all points accessible from w; hence w is not separable from these points by a neighborhood.

Let us give a completeness proof for $X = [0, 1]$, the unit interval of the real line. Recall that open sets in $[0, 1]$ are finite or a countable unions of open intervals.

THEOREM 9.2. *A formula is derivable in NJp iff it is valid in* $[0, 1]$.

Proof. Derivable formulas are valid by the previous Theorem. The proof of the opposite implication takes the rest of the present subsection. Fix a formula α and a finite pointed Kripke model $(0, \mathbf{K}, R, \models)$ [Theorem 8.2.(b)] such that:

$$\mathbf{K} =< 0, \ldots, N-1 >, \ 0 \not\models \alpha$$

and R is a *partial ordering* on $\mathbf{K}$.

If $N = 1$, let

$$V(p) = \ if \ 0 \models p \ then \ [0, 1] \ else \ \emptyset$$

This relation extends to all formulas: By induction on α we prove:

$$V(\alpha) = \ if \ 0 \models \alpha \ then \ [0, 1] \ else \ \emptyset,$$

since computations follow classical truth tables: $Interior(Y) = Y$ for $Y = [0, 1], \emptyset$. In particular, $V(\alpha) = \emptyset$; that is, α is refuted in $[0, 1]$. This concludes the case $N = 1$.

From now on, assume that $N > 1$.

Consider the set of all infinite sequences of worlds (non-strictly) monotonic with respect to R:

$$N^{mon} =_{def} \{s : \omega \to \mathbf{K} \mid \quad \forall k\ (s(k) = s(k+1) \vee\ R(s(k))s(k+1))\}$$

In fact, this set already provides a model in a compact space similar to Cantor set if we use the standard metric:

$$d(s,t) = 2^{-k}, \qquad where\ k = \text{ the least } n \text{ such that } (s(n) \neq t(n))$$

We shall not use this fact to imbed N^{mon} into [0,1].

Note. Every sequence $s \in N^{mon}$ stabilizes (after at most $N-1$ jumps).

Let $n(s)$ be the stabilization moment for s, that is the least natural number satisfying the condition:

$$(\forall m \geq n(s))\ s(m) = s(n(s)),$$

and let

$$w(s) = s(n(s))$$

be the stable value.

Define $q := 1/(2N-1)$ and note that:

$$q \leq 1/3 < 1, \qquad \frac{1}{1-q} = \frac{2N-1}{2(N-1)} = \frac{1}{2(N-1)q}$$

Define for all $s, t : \omega \to \mathbf{K}$:

$$\tilde{s} := \sum_{n=0}^{\infty} 2s(n)q^{n+1}$$

$$s|_k := \sum_{n=0}^{k} 2s(n)q^{n+1} \qquad s =_k t := (\forall n \leq k)s(n) = t(n)$$

From now on we assume $Rww' \to w \leq w'$. In other word, the worlds are enumerated from the root by "horisontal slices". In this Chapter s, t, u stand for elements of N^{mon}, and x, y, z stand for real numbers in $[0, 1]$.

LEMMA 9.1. *For all* $s, t \in N^{mon}$*:*

(a) $$0 \leq \tilde{s} \leq 1;$$

(b) $$s|_k \leq \tilde{s} \leq s|_k + q^{k+1},$$

with the first $\leq$ *replaced by* $<$ *if* $s(k) \neq 0$*;*

(c) *if* $s =_k t$*, then* $|\tilde{s} - \tilde{t}| \leq q^{k+1}$

with $\leq$ *replaced by* $<$ *if* $s(k) \neq 0$*;*

(d) *if* $s \neq_k t$*, then* $|\tilde{s} - \tilde{t}| > q^{k+1}$.

Proof. Part (a):

$$0 \leq \tilde{s} = \sum_{n=0}^{\infty} 2s(n)q^{n+1} \leq 2(N-1)\sum_{n=0}^{\infty} q^{n+1} = 2(N-1)q/1-q = \frac{2(N-1)q}{2(N-1)q} = 1$$

Part (b): All terms in the series for $\tilde{s}$ are non-negative. If $s(k) \neq 0$, then $s(l) \neq 0$ for $l \geq k$, since $s \in N^{mon}$ and R is a partial ordering. Hence similarly to Part (a):

$$s|_k \leq \tilde{s} = s|_k + \sum_{n=k+1}^{\infty} 2s(n)q^{n+1} \leq s|_k + 2(N-1)\sum_{n=k+1}^{\infty} q^{n+1}$$

$$= s|_k + 2(N-1)q^{k+2}/1-q = s|_k + q^{k+1}$$

Part (c): $s =_k t$ implies $s|_k = t|_k$ and by Part (b), $s|_k \leq \tilde{t} \leq s|_k + q^{k+1}$. Hence $|\tilde{s} - \tilde{t}| \leq q^{k+1}$. Here we have the following situation:

```
|------    |-------    |-------    |
s|k        s~          t~          s|k + q^{k+1}
```

Part (d): It is enough to prove (d) for the minimal number k satisfying $s \neq_k t$, since q^k decreases with k. In other words, we assume $s =_{k-1} t$ and (choosing one of two possibilities) $t(k) > s(k)$. Then $t(k) \geq s(k) + 1 > 0$, and by Part (b):

$$\tilde{t} > t|_k = t|_{k-1} + 2t(k)q^{k+1} \geq t|_{k-1} + 2(s(k)+1)q^{k+1} =$$

$$s|_{k-1} + 2s(k)q^{k+1} + 2q^{k+1} = s|_k + 2q^{k+1}.$$

Together with $\tilde{s} \leq s|_k + q^{k+1}$, this implies $\tilde{t} > \tilde{s} + q^{k+1}$. ⊣

LEMMA 9.2. *(a) If* $m \geq n(s) - 1$, *then*

$$\tilde{s} = s|_m + \frac{w(s)}{N-1}q^{m+1}$$

In particular,

$$\tilde{s} = s|_{n(s)-1} + \frac{w(s)}{N-1}q^{n(s)}$$

(b) If $s \neq t$, $k =$ *the least* n $(s(n) \neq t(n))$, *then*

$$t(k) > s(k) \textit{ implies } \tilde{t} > \tilde{s} + q^{k+1}$$

Proof. Part (a):

$$\tilde{s} = s|_m + \sum_{n=m+1}^{\infty} 2s(n)q^{n+1} = s|_m + 2w(s)\sum_{n=m+1}^{\infty} q^{n+1} =$$

$$s|_n + 2w(s)q^{m+2}/(1-q) = s|_m + \frac{w(s)}{N-1}q^{m+1}$$

Part (b): We have $t(k) \geq s(k) + 1 > 0$ and

$$\tilde{t} > t|_k = t|_{k-1} + 2t(k)q^{k+1} \geq t|_{k-1} + 2(s(k)+1)q^{k+1} =$$
$$s|_{k-1} + 2s(k)q^{k+1} + 2q^{k+1} = s|_k + 2q^{k+1} \geq \tilde{s} + q^{k+1}$$

⊣

LEMMA 9.3. *If $s, t \in N^{mon}$ and $s =_{n(t)} t$, then $Rw(t)w(s)$.*

Proof. If $n(t) < n(s)$, then:
$Rs(n(t))s(n(s))$ [monotonicity of s] $\leftrightarrow Rt(n(t))s(n(s))$ [since $s =_{n(t)} t$] $\leftrightarrow Rw(t)w(s)$.

If $n(t) \geq n(s)$, then $w(t) = t(n(t)) = s(n(t))$ [since $s =_{n(t)} t$] $= s(n(s))$ [monotonicity] $= w(s)$. ⊣

For $x \in [0,1]$ define

$$s[x] := \begin{cases} \lambda n.0, & \text{if } |x - \tilde{s}| \geq q^{n(s)+2} \text{ for every } s \in N^{mon} \\ \text{the } s \in N^{mon} & \text{closest to } x \text{ such that } |x - \tilde{s}| < q^{n(s)+2} \text{ otherwise} \end{cases}$$

The latter condition means $|x - \tilde{s}| < |x - \tilde{t}|$ for all $t \in N^{mon}$.

LEMMA 9.4. *$s[x]$ is defined.*

Proof. Assume that s, t both satisfy the last clause in the definition of $s[x]$, that is,

$$q^{n(s)+2}, q^{n(t)+2} > |x - \tilde{t}| = |x - \tilde{s}| = \frac{\tilde{t} - \tilde{s}}{2}, \tag{9.4}$$

and $\tilde{s} < \tilde{t}$. Then $\tilde{s} < x < \tilde{t}$ [otherwise one of $\tilde{s}, \tilde{t}$ is closer to x than the other]. We have $n(s) \leq n(t)$, since otherwise $|\tilde{s} - \tilde{t}| < q^{n(s)+2} + q^{n(t)+2} < 2q^{n(t)+2} < q^{n(t)+1}$, $s =_{n(t)} t$, $Rw(t)w(s)$ and $\tilde{t} \leq \tilde{s}$. This implies $|\tilde{s} - \tilde{t}| < q^{n(s)+1}$, $s|_{n(s)} = t|_{n(s)}$, $Rw(s)w(t)$, $w(s) \leq w(t)$. Together with $\tilde{s} < \tilde{t}$ this implies $w(s) < w(t)$, $n(s) < n(t)$. Since $s \neq_{n(t)} t$, we have $\tilde{t} > \tilde{s} + q^{n(t)+1}$. Together with (9.4) this implies $q^{n(t)+2} > \frac{q^{n(t)+1}}{2}$ and $q > 1/2$, a contradiction. ⊣

Let

$$w(x) := w(s[x])$$

Since $s[\tilde{t}] = t$ for $t \in N^{mon}$, we have $w(\tilde{t}) = w(t)$. We also write $\tilde{s}[x]$ for $\widetilde{s[x]}$.

LEMMA 9.5. *If $|x - \tilde{t}| < q^{n(t)+2}$, then $Rw(t)w(x)$*

Proof. From $|x - \tilde{t}| < q^{n(t)+2}$, we have $|x - \tilde{s}[x]| \leq |x - \tilde{t}| < q^{n(t)+2}$. Then $|\tilde{t} - \tilde{s}[x]| < 2q^{n(t)+2} < q^{n(t)+1}$, implying $Rw(t)w(s[x])$, that is, $Rw(t)w(x)$. ⊣

For every propositional variable p define a set $V(p)$ by

$$x \in V(p) \leftrightarrow w(x) \models p \tag{9.5}$$

where $\models$ stands for truth in the model $(0, \mathbf{K}, R)$.

LEMMA 9.6. *The set $V(p)$ is open*

Proof. Assume $x \in V(p)$. If $w(x) = 0$, then $0 \models p$ and $V(p) = [0, 1]$. Assume now $w(x) \neq 0$ (which implies that $s[x]$ is computed by the second clause of the definition) and let ϵ be the distance from x to the closest end of the interval $\tilde{s}[x] \pm q^{n(s[x])+2}$:

$$\epsilon := min(x - (\tilde{s}[x] - q^{n(s[x])+2}), \tilde{s}[x] + q^{n(s[x])+2} - x)$$

Then $\epsilon > 0$. Take an $y \in [0, 1]$ with $|x - y| < \epsilon$. To prove $w(y) \models p$, note that $|y - \tilde{s}[x]| < q^{n(s[x])+2}$. This implies $|y - \tilde{s}[y]| \leq |y - \tilde{s}[x]|$ and $|\tilde{s}[y] - \tilde{s}[x]| < q^{n(s[x])+1}$. Hence $Rw(x)w(y)$ and $w(y) \models p$. ⊣

THEOREM 9.3. *For arbitrary formula α*

$$x \in V(\alpha) \leftrightarrow w(x) \models \alpha \tag{9.6}$$

Proof. We use induction on α. The induction base is the relation (9.5). Induction step for $\alpha \equiv (\gamma \odot \beta)$, where $\odot \equiv \&, \vee$:

$$x \in V((\gamma \odot \beta)) \;\leftrightarrow\; (x \in V(\gamma)) \odot (x \in V(\beta)) \;\leftrightarrow\; w(x) \models \gamma \odot w(x) \models \beta$$

$$\leftrightarrow\; w(x) \models (\gamma \odot \beta),$$

as required for (9.6).

Let $\alpha \equiv (\gamma \rightarrow \beta)$. Recall that:

$$x \in V(\gamma \rightarrow \beta) \;\leftrightarrow\; x \in Interior(([0, 1] - V(\gamma)) \cup V(\beta)) \;\leftrightarrow\; (\exists \epsilon > 0)$$

$$(\forall y \in [0, 1])(|y - x| < \epsilon \rightarrow (y \in V(\gamma) \textit{ implies } y \in V(\beta))) \tag{9.7}$$

Assume first $x \in V(\alpha)$. Take an ϵ satisfying (9.7) and an arbitrary w with $Rw(x)w$, $w \models \gamma$. Take a natural number $n_0 \geq n(s[x]) + 2$ such that $q^{n_0+1} < \epsilon$. Define a sequence t by:

$$t(n) := \textbf{if } n \leq n_0 \textbf{ then } s[x](n) \textbf{ else } w$$

Since $s[x](n_0) = t(n_0) = w(x)$, we have $Rt(n_0)w$ and $t \in N^{mon}$, as well as $w(t) = w \models \gamma$, $s[x] =_{n_0} t$ and $|x - \tilde{t}| \leq q^{n_0+1} < \epsilon$.

Hence $\tilde{t} \in V(\gamma)$, so $\tilde{t} \in V(\beta)$, and $w = w(t) \models \beta$ by IH. This implies $w(x) \models \alpha$.

Now assume that $w(x) \models (\gamma \rightarrow \beta)$, but $x \notin V(\gamma \rightarrow \beta)$. Take a sequence $x_n \in V(\gamma) - V(\beta)$, $x_n \rightarrow_{n \rightarrow \infty} x$. If $w(x) = 0$ then $Rw(x)w(x_m)$ for all m. Otherwise, $|x - \tilde{s}[x]| < q^{n(s[x])+2}$, and for sufficiently big m, $|x_m - \tilde{s}[x]| < q^{n(s[x])+2}$, hence $Rw(x)w(x_m)$. By IH, $w(x_m) \models \gamma$, and by $Rw(x)w(x_m)$, $w(x_m) \models \beta$. Again by IH, $x_m \in V(\beta)$, a contradiction. ⊣

To finish the proof of the topological completeness for $[0, 1]$, recall that $0 \not\models \alpha$. Then for the constant 0-sequence $\mathbf{0}$, we have $\tilde{\mathbf{0}} \notin V(\alpha)$; hence $V(\alpha) \neq [0, 1]$ as required. ⊣

Chapter 10

Proof-Search

10.1. Tableaux: System LJpm*

For each sequent S, we describe a tree Tr_S obtained from S by a bottom-up application of the inference rules of the system LJpm. If the goal sequent S is, say, of the form:

$$\varphi\&\psi, \Gamma \Rightarrow \Delta,$$

then it can be derived from a simpler sequent:

$$\varphi, \psi, \Gamma \Rightarrow \Delta$$

by the rule $(\Rightarrow \&)$. Therefore the first step of the construction of Tr_S can be

$$\frac{\varphi, \psi, \Gamma \Rightarrow \Delta}{\varphi\&\psi, \Gamma \Rightarrow \Delta.} \tag{10.1}$$

In this way Tr_S is extended upward with the ultimate goal of obtaining a derivation of S. Bottom-up applications of two-premise rules lead to a branching of Tr_S. If, say, $\psi = \psi_1 \vee \psi_2$, then (10.1) can be extended as follows:

$$\cfrac{\cfrac{\varphi, \psi_1, \Gamma \Rightarrow \Delta \quad \varphi, \psi_2, \Gamma \Rightarrow \Delta}{\varphi, \psi, \Gamma \Rightarrow \Delta}}{\varphi\&\psi, \Gamma \Rightarrow \Delta}$$

This kind of branching is called *conjunctive* or *and*-branching: To prove a final goal sequent, it is necessary to prove *both* subgoals. In this way we can treat all invertible rules of the system LJpm: If a potential principal formula of such a rule is present, it is eventually analyzed. Non-invertible rules present a problem and will cause an extension of the language. Consider a sequent:

$$\neg p \to q \vee r \Rightarrow \neg p \to q, \neg p \to r, \neg p$$

The only invertible rule applicable here (bottom-up) is $(\to\Rightarrow)$, but it is redundant, since $\neg p$ is already present in the succedent. Hence we must try non-invertible rules. To make the search exhaustive, all of these rules are to be tried in parallel. The proof-search is successful iff at least one of these parallel processes terminates. This situation is called *disjunctive* or *or*-branching.

To describe disjunctive branching, we introduce more complicated objects: Finite ordered sequences or lists of sequents, written as follows:

$$S_1 * S_2 * \ldots * S_n \quad (n \geq 1) \tag{10.2}$$

These are called *tableaux*. The order of *components* $S_1, \ldots, S_n$ is in fact irrelevant, but we keep it fixed to facilitate notation. Tableaux are derivable objects of our new system LJpm*. Its rules are the rules of LJpm applied to one of the components S_i. Note that the rule $\Rightarrow\rightarrow$ adds a new component (when viewed bottom-up) and hence turns a sequent into a tableau.

System LJpm*.
Axioms:

$$T * \varphi, \Gamma \Rightarrow \Delta, \varphi * T' \qquad\qquad T * \bot, \Gamma \Rightarrow \Delta * T'$$

Inference rules:

$$\frac{T * \Gamma \Rightarrow \Delta, \varphi * T' \quad T * \Gamma \Rightarrow \Delta, \psi * T'}{T * \Gamma \Rightarrow \Delta, \varphi \& \psi * T'} \qquad \frac{T * \varphi, \psi, \Gamma \Rightarrow \Delta * T'}{T * \varphi \& \psi, \Gamma \Rightarrow \Delta * T'}$$

$$\frac{T * \varphi, \Gamma \Rightarrow \Delta * T' \quad T * \psi, \Gamma \Rightarrow \Delta * T'}{T * \varphi \vee \psi, \Gamma \Rightarrow \Delta * T'} \qquad \frac{T * \Gamma \Rightarrow \Delta, \varphi, \psi * T'}{T * \Gamma \Rightarrow \Delta, \varphi \vee \psi * T'}$$

$$\frac{T * \varphi \rightarrow \psi, \Gamma \Rightarrow \Delta, \varphi * T' \quad T * \psi, \Gamma \Rightarrow \Delta * T'}{T * \varphi \rightarrow \psi, \Gamma \Rightarrow \Delta * T'}$$

$$\frac{T * \Gamma \Rightarrow \Delta, \varphi \rightarrow \psi * \varphi, \Gamma \Rightarrow \psi * T'}{T * \Gamma \Rightarrow \Delta, \varphi \rightarrow \psi * T'}$$

All the rules of LJpm* are in fact invertible, but we call $\Rightarrow\rightarrow$ *quasi-invertible* to stress its connection with the non-nvertible rule $\Rightarrow\rightarrow$ of LJpm.

Let us prove first that the new system is equivalent to the old one. We could just interpret $*$ as a disjunction, but this would introduce cuts.

LEMMA 10.1. *Any derivation of a tableau* $S_1 * S_2 * \ldots * S_n$ *in* LJpm* *can be pruned into a derivation of one* S_i *in LJpm by deleting some components of tableaux.*

Proof. Use induction on the height of the derivation. The axioms of LJpm* are pruned into the corresponding axioms of LJpm, and all remaining components (that is, T, T') are deleted.

If some non-active component (occurring in T, T') is retained in at least one of the premises of a rule, it is also retained in the conclusion. For example, the rule:

$$\frac{\varphi, \Gamma \Rightarrow \Delta * T * S \quad \psi, \Gamma \Rightarrow \Delta * T * S}{\varphi \vee \psi, \Gamma \Rightarrow \Delta * T * S}$$

is transformed into the repetition:

$$\frac{S}{S}$$

if S were retained in one of the premisses. Otherwise only the active sequents are retained, for example:

$$\frac{\varphi, \Gamma \Rightarrow \Delta * T \qquad \psi, \Gamma \Rightarrow \Delta * T}{\varphi \vee \psi, \Gamma \Rightarrow \Delta * T}$$

is transformed into:

$$\frac{\varphi, \Gamma \Rightarrow \Delta \qquad \psi, \Gamma \Rightarrow \Delta}{\varphi \vee \psi, \Gamma \Rightarrow \Delta}$$

if the leftmost sequents were retained in both premises. ⊣

THEOREM 10.1. *System LJpm* is equivalent to* LJpm*: A sequent is derivable in* LJpm* *iff it is derivable in* LJpm.

Proof.One direction is the previous Lemma. To prove the other direction we add redundant sequents to transform LJpm-derivations into a LJpm* derivation. More precisely, we proceed by induction on LJpm-derivations. The basis case is obvious. The induction step is proved by cases depending on the last rule L in the derivation. When L is invertible, the induction hypothesis is used. Consider a non-invertible rule $(\Rightarrow\rightarrow)$:

$$\frac{\varphi, \Gamma \Rightarrow \psi}{\Gamma \Rightarrow \Delta, \varphi \rightarrow \psi}$$

By the induction hypothesis, we have a derivation d of $\varphi, \Gamma \Rightarrow \psi$ in LJpm*. Prefixing $\Gamma \Rightarrow \Delta, (\varphi \rightarrow \psi)*$ to all tableaux in d, we obtain a derivation of $\Gamma \Rightarrow \Delta, \varphi \rightarrow \psi * \varphi, \Gamma \Rightarrow \psi$ in LJpm* and complete the derivation by applying the rule $(\Rightarrow\rightarrow)$ of LJpm*. ⊣

10.2. Proof-Search Procedure

Let us describe a *proof-search procedure* for the system LJpm* that constructs a proof-search tree Tr_{T_0} for each tableau T_0. It works in a bottom-up way: For a goal tableau T, it either declares that T cannot be analyzed further, or it analyzes T to determine a new goal T_1 or two new goals T_1 and T_2, such that T is obtained from these by some rule L. We say that the proof-search tree is *extended by using L bottom-up*:

$$\frac{T_1}{T}\ L \qquad \text{or} \qquad \frac{T_1 \qquad T_2}{T}\ L$$

At each stage except possibly the final one, the proof-search tree Tr_{T_0} has a finite list of *goals* (leaf nodes of the tree). The *tree extension* step consists in picking one of the goals T (for example, the leftmost one) to analyze. Then T

is replaced in the list of goals by T_1 or by T_1 and T_2. If T is an axiom, it is not analyzed but simply deleted from the list of goals and declared a *closed* leaf of the tree. Then the extension step is repeated. If T is not an axiom and cannot be analyzed further, then it is declared a *terminal node*, the proof-search process stops, and the initial tableau T_0 is declared underivable. If the list of goals is empty, that is, all leaf nodes of the tree are closed, then the tree is *closed*, and the initial tableau is derivable. (Indeed in this case the tree Tr_{T_0} proceeds from axioms by the rules of LJpm*.)

At the initial step, the whole tree consists of exactly one node, namely, the tableau T_0 we are testing for derivability.

Before presenting a proof of the termination and completeness of this proof-search procedure, consider two examples. We preserve principal formulas in all rules (see the beginning of Section 10.3. below).

EXAMPLE 10.1. Consider a sequent $\neg(p \vee \neg p) \Rightarrow \bot$.

We omit the right premise $\bot, \Gamma \Rightarrow \Delta$ (axiom) in the rule $\neg \Rightarrow$ and some occurrences of $\bot$ in the succedent, since these do not contribute to proof-search:

$$\dfrac{\dfrac{\dfrac{\dfrac{\dfrac{\neg(p \vee \neg p) \Rightarrow p \vee \neg p, p, \neg p * \neg(p \vee \neg p), p \Rightarrow p \vee \neg p, p, \neg p}{\neg(p \vee \neg p) \Rightarrow p \vee \neg p, p, \neg p * \neg(p \vee \neg p), p \Rightarrow p \vee \neg p}}{\neg(p \vee \neg p) \Rightarrow p \vee \neg p, p, \neg p * \neg(p \vee \neg p), p \Rightarrow \bot}}{\neg(p \vee \neg p) \Rightarrow p \vee \neg p, p, \neg p}}{\neg(p \vee \neg p) \Rightarrow p \vee \neg p}}{\neg(p \vee \neg p) \Rightarrow \bot}$$

Our sequent is derivable.

EXAMPLE 10.2. Consider a proof-search tree for a sequent $\neg(p\&q) \Rightarrow \neg p \vee \neg q$:

$$\dfrac{\dfrac{\dfrac{\neg(p\&q) \Rightarrow \neg p \vee \neg q, \neg p, \neg q, p, p\&q \qquad \neg(p\&q) \Rightarrow \neg p \vee \neg q, \neg p, \neg q, q, p\&q}{\neg(p\&q) \Rightarrow \neg p \vee \neg q, \neg p, \neg q, p\&q}}{\neg(p\&q) \Rightarrow \neg p \vee \neg q, \neg p, \neg q}}{\neg(p\&q) \Rightarrow \neg p \vee \neg q}$$

Let:

$$S := \neg(p\&q) \Rightarrow \neg p \vee \neg q, \neg p, \neg q, p, p\&q;$$

$$S_p := \neg(p\&q), p \Rightarrow p\&q, \quad \text{and} \quad S_q := \neg(p\&q), q \Rightarrow p\&q.$$

Continue the search up the left branch:

$$\dfrac{\dfrac{\dfrac{S * \neg(p\&q), p \Rightarrow p\&q, p * S_q \qquad \dfrac{T_1 \quad T_2}{S * \neg(p\&q), p \Rightarrow q * S_q}}{S * S_p * S_q}}{S * \neg(p\&q), p \Rightarrow * \neg(p\&q), q \Rightarrow}}{\neg(p\&q) \Rightarrow \neg p \vee \neg q, \neg p, \neg q, p, p\&q}$$

where:

$$T_1 := S \;*\neg(p\&q), p \Rightarrow p\&q, q \;*\neg(p\&q), q \Rightarrow p\&q, p$$

and $T_2 := S * \neg(p\&q), p \Rightarrow p\&q, q * \neg(p\&q), q \Rightarrow p\&q, q$.

The tableau T_1 is terminal (no rule can be meaningfully applied bottom-up to it). It produces the following refuting Kripke model for the original sequent $\neg(p\&q) \Rightarrow \neg p \vee \neg q$, where $\mathbf{0}, \mathbf{1}, \mathbf{2}$ correspond to components of T_1:

$$S; \qquad \neg(p\&q), p \Rightarrow p\&q, q; \qquad \neg(p\&q), q \Rightarrow p\&q, p$$

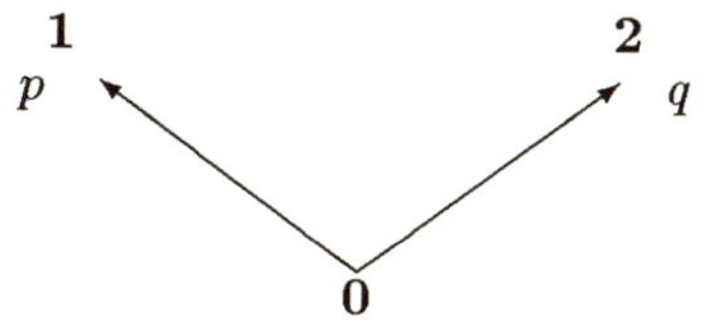

Indeed the antecedent is true in the world $\mathbf{0}$: $V(\neg(p\&q),\mathbf{0}) = 1$ because $p\&q$ is refuted in all three worlds: In $\mathbf{0}$ and $\mathbf{1}$, since q is false there and in $\mathbf{2}$, since p is false there. On the other hand $\neg p, \neg q$ are both false in the world $\mathbf{0}$; $V(\neg p,\mathbf{0}) = 0$, since $V(p, \mathbf{1}) = 1$; and $V(\neg q,\mathbf{0}) = 0$, since $V(q, \mathbf{2}) = 1$. Hence $V(\neg p \vee \neg q,\mathbf{0}) = 0$ as required.

10.3. Complete Proof-Search Strategy

Let us impose some restrictions on tree extension steps. Structural rules are used only as parts in a step by a logical rule, and are not mentioned separately. Suitable sequents of maximal complexity are analyzed first. To avoid repeated analyzis of the same formula, all principal formulas of invertible rules are preserved; that is, every invertible step is always preceded by a contraction of the principal formula. For example, $\& \Rightarrow, \Rightarrow \vee$ and $\rightarrow\Rightarrow$ take the following form:

$$\frac{\dfrac{T_1 *\; \alpha, \beta, \alpha\&\beta, \Gamma \Rightarrow \Delta *\; T_2}{T_1 *\; \alpha\&\beta, \alpha\&\beta, \Gamma \Rightarrow \Delta *\; T_2}}{T_1 *\; \alpha\&\beta, \Gamma \Rightarrow \Delta *\; T_2} \qquad \frac{\dfrac{T_1 *\; \Gamma \Rightarrow \Delta, \alpha \vee \beta, \alpha, \beta *\; T_2}{T_1 *\; \Gamma \Rightarrow \Delta, \alpha \vee \beta, \alpha \vee \beta *\; T_2}}{T_1 *\; \Gamma \Rightarrow \Delta, \alpha \vee \beta *\; T_2}$$

$$\frac{\dfrac{\dfrac{T_1 *\; \alpha \rightarrow \beta, \Gamma \Rightarrow \Delta, \alpha *\; T_2}{T_1 *\; \alpha \rightarrow \beta, \alpha \rightarrow \beta, \Gamma \Rightarrow \Delta, \alpha, *\, T_2} \quad T_1 *\; \beta, \alpha \rightarrow \beta, \Gamma \Rightarrow \Delta *\; T_2}{T_1 *\; \alpha \rightarrow \beta, \alpha \rightarrow \beta, \Gamma \Rightarrow \Delta *\; T_2}}{T_1 *\; \alpha \rightarrow \beta, \Gamma \Rightarrow \Delta *\; T_2}$$

Consider the following *proof strategy* avoiding redundancy.

DEFINITION 10.1. *A sequent* $\Gamma \Rightarrow \Delta$ subsumes *a sequent* $\Gamma' \Rightarrow \Delta'$ *(and* $\Gamma' \Rightarrow \Delta'$ is subsumed by $\Gamma \Rightarrow \Delta$*) iff* $\Gamma \subseteq \Gamma'$ *and* $\Delta \subseteq \Delta'$ *as sets of formulas. (Note that* $(\&\Gamma \rightarrow \vee\Delta) \vdash \&\Gamma' \rightarrow \vee\Delta'$ *in this case).*

An extension step applied to a tableau T is admissible *if none of the new components S' (there is just one S' for one-premise rule) subsumes any component of any tableau from T down the root (original sequent) of the proof-search tree.*

LEMMA 10.2. *If principal formulas of invertible rules are preserved and only admissible tree extension steps are applied, then the proof-search in* LJpm *terminates.*

Proof. Every sequent in the proof-search tree for a sequent S consists of subformulas of S; hence there is only a finite number N of different sequents (when antecedents and succedents are treated as sets of formulas). Every admissible extension step adds a new sequent to a given branch of the proof-search tree, so the length of the branch is bounded by N. Since the tree is binary (it has at most two branches at each node), it is finite. ⊣

LEMMA 10.3. *If principal formulas of invertible rules are preserved, only admissible tree extension steps are applied, and T is a terminal node of a proof-search tree (no extension step is admissible), then T can be pruned (by deleting components properly subsumed by others) into a saturated set T^- of sequents.*

Proof. Observe that if a tableau T is situated over some tableau T_1 in some branch of the proof-search tree (that is, T is obtained from T_1 by a series of extension steps), then every component of T_1 subsumes some component of T. Indeed if a component of the conclusion of a rule is changed in the passage to the premise, then this rule is invertible, and the component is just extended.

Let T be a terminal node. Delete all components of T that properly subsume other components and prove that the resulting set T^- of sequents is saturated. Let $S \in T^-$. Then S is saturated for invertible rules: For any potential bottom-up application of an invertible rule to a formula in S, one of the newly generated components coincides with S. Otherwise this component S' properly contains S. Since T is a terminal node, S' subsumes one of the components, say S'', in the branch from T down to the root. As previously observed, S'' subsumes some component in T, and hence S properly subsumes some component of T, that is, $S \notin T^-$.

To prove that T is saturated for non-invertible rules (of LJpm) to formulas in a given component $S \in T^-$, consider the new component S' to be generated by such an application. As in the case of an invertible rule, a terminal tableau T contains a component subsumed by S'; that is, it is saturated. ⊣

It remains to describe the construction of a refuting model for the original tableau S_0 from the (non-closed) terminal tableau T in the proof-search tree Tr_{T_0}. By the previous Lemma 10.3., a subset T^- of T is saturated. By Theorem 8.2., it produces a Kripke model $K \equiv < W, M, V >$ falsifying all sequents in T^- by (8.8). Since every sequent in $T - T^-$ subsumes some sequent in T^-, it is also falsified in the same world of W.

LEMMA 10.4. *Let T be a terminal tableau (leaf node) in a proof-search tree Tr_{T_0} for a tableau:*

$$T_0 \equiv S_1 * \ldots * S_n,$$

and let $K \equiv < W, M, V >$ be a Kripke model such that $w_i \in W$ falsifies the ith component of T. Then w_i falsifies S_i for every $i \leq n$.

Proof. Easy induction on height of Tr_{T_0}. Induction base: $\mathrm{Tr}_{T_0} = T$. Apply the assumption.

Induction step. The T_0 is derived by a rule L applied bottom-up to its component S_I, $1 \leq I \leq n$. Let T' be the premise of L in the branch of Tr_{T_0} leading to node T:

$$\begin{array}{c} T \\ \vdots \\ \dfrac{T' \quad (T'')}{T_0} \end{array}$$

If L is invertible, that is, $L \not\equiv (\Rightarrow\rightarrow)$, then T' is obtained from T_0 by replacing S_I by S'_I. By IH, w_i falsifies S_i for $i \neq I$, and w_I falsifies S'_I. Since L is invertible, S_I implies S'_I; hence w_I falsifies S_I.

If $L \equiv (\Rightarrow\rightarrow)$, then $T' \equiv T * S'$, and IH is applicable. ⊣

THEOREM 10.2. *(a) The proof-search procedure is sound and complete.*

(b) Intuitionistic propositional logic is decidable

Proof. If the proof-search does not terminate in a derivation, take a terminal tableau and apply Lemma 10.3. to construct a refuting model. ⊣

Chapter 11

System LJp

One-succedent version LJp of LJpm is obtained by restricting the succedent to one formula in LJpm. In other words, sequents are of the form $\Gamma \Rightarrow \alpha$, and the rules are as follows:

System LJp Axioms:

$$\phi \Rightarrow \phi \qquad\qquad \bot \Rightarrow \phi$$

Inference rules:

$$\frac{\Gamma \Rightarrow \phi \quad \Gamma \Rightarrow \psi}{\Gamma \Rightarrow \phi \& \psi} \qquad \frac{\phi, \psi, \Gamma \Rightarrow \theta}{\phi \& \psi, \Gamma \Rightarrow \theta}$$

$$\frac{\phi, \Gamma \Rightarrow \theta \quad \psi, \Gamma \Rightarrow \theta}{\phi \vee \psi, \Gamma \Rightarrow \theta} \qquad \frac{\Gamma \Rightarrow \phi_i}{\Gamma \Rightarrow \phi_0 \vee \phi_1}$$

$$\frac{\phi \rightarrow \psi, \Gamma \Rightarrow \phi \quad \psi, \Gamma \Rightarrow \theta}{\phi \rightarrow \psi, \Gamma \Rightarrow \theta} \qquad \frac{\phi, \Gamma \Rightarrow \psi}{\Gamma \Rightarrow \phi \rightarrow \psi}$$

$$\frac{\phi, \phi, \Gamma \Rightarrow \Delta}{\phi, \Gamma \Rightarrow \Delta}\ contr \qquad \frac{\Gamma \Rightarrow \Delta}{\phi, \Gamma \Rightarrow \Delta}\ weak$$

The following rules are derivable:

$$\frac{\phi, \Gamma \Rightarrow \bot}{\Gamma \Rightarrow \neg\phi} \qquad \frac{\neg\phi, \Gamma \Rightarrow \phi}{\neg\phi, \Gamma \Rightarrow \theta}$$

11.1. Translating LJpm into LJp

11.2. A Disjunctive translation

For $\Delta \equiv \alpha_1, \ldots, \alpha_n$ let $\vee\Delta := \alpha_1 \vee \ldots \vee \alpha_n$.

LEMMA 11.1. *(a)* $\Gamma \Rightarrow \Delta$ *is derivable in* $LJpm$ *iff* $\Gamma \Rightarrow \vee\Delta$ *is derivable in* $LJp + cut$.

(b) In particular $LJpm$ *is equivalent to* $LJp + cut$ *for sequents* $\Gamma \Rightarrow \alpha$.

Proof. Part (a): The translation of every rule of $LJpm$ is derivable in $LJp+$ cut. For example:

$$\frac{\Gamma \Rightarrow \Delta, \alpha \quad \Gamma \Rightarrow \Delta, \beta}{\Gamma \Rightarrow \Delta, \alpha\&\beta}$$

is translated as:

$$\frac{\dfrac{\Gamma \Rightarrow \delta \vee \alpha \quad \Gamma \Rightarrow \delta \vee \beta}{\Gamma \Rightarrow (\delta \vee \alpha)\&(\delta \vee \beta)} \quad d:\ (\delta \vee \alpha)\&(\delta \vee \beta) \Rightarrow \delta \vee (\alpha\&\beta)}{\Gamma \Rightarrow \delta \vee (\alpha\&\beta)},$$

where (disregarding some *weak, contr*):

$$\frac{\dfrac{\dfrac{\delta \Rightarrow \delta}{\delta \Rightarrow \delta \vee (\alpha\&\beta)} \quad \dfrac{\dfrac{\delta \Rightarrow \delta}{\delta \Rightarrow \delta \vee (\alpha\&\beta)} \quad \dfrac{\alpha \Rightarrow \alpha \quad \beta \Rightarrow \beta}{\alpha, \beta \Rightarrow \alpha\&\beta}}{\alpha, (\delta \vee \beta) \Rightarrow \delta \vee (\alpha\&\beta)}}{(\delta \vee \alpha), (\delta \vee \beta) \Rightarrow \delta \vee (\alpha\&\beta)}}{d:\ (\delta \vee \alpha)\&(\delta \vee \beta) \Rightarrow \delta \vee (\alpha\&\beta)}$$

In the opposite direction, a derivation in LJp is already a derivation in LJpm (up to structural rules), and the passage from $\Gamma \Rightarrow \vee\Delta$ to $\Gamma \Rightarrow \Delta$ is done by cut with $\vee\Delta \Rightarrow \Delta$. ⊣

11.3. Pruning, Permutability of Rules

The translation of LJpm into LJp given in the previous Section 11.2. distorts the structure of derivations and introduces cuts. This may be inconvenient if the goal is for example to extract a program.

In the present Section we describe a translation that preservs the structure of a derivation better and prunes multiple-succedent sequents into *1-sequents*, that is, sequents of the form $\Gamma \Rightarrow \alpha$. The translation uses permutation of inferences along the lines of [11].

Let us first describe pruning transformations that delete redundant parts of a deduction.

DEFINITION 11.1. *A derivation in* $LJpm$ *is* pruned *if weakenings occur only as follows (with only other weakenings intervening):*

(a) immediately preceding the endsequent;

(b) to introduce the side formula of $\Rightarrow\rightarrow$, *the other one not being introduced by a weakening;*

(c) to introduce one of the parametric formulas into one premise of a two-premise inference, this formula not being introduced into the other premise by a weakening.

LEMMA 11.2. *Every derivation $d : \Gamma \Rightarrow \Delta$ in LJpm can be transformed into a pruned derivation $d' : \Gamma \Rightarrow \Delta$ by moving weakenings downward and deleting occurrences of formulas and whole branches.*

Proof. We use induction on d. The induction base is obvious; for the induction step, consider the last rule L in d.

Case 1. The L is a two-premise rule. If at least one of the side formulas of L is introduced by weakening, then L is deleted together with the whole branch ending in the other premise; for example:

$$\dfrac{\dfrac{e:\ \Gamma' \Rightarrow \Delta'}{\Gamma \Rightarrow \Delta, \alpha}\,weak \qquad \Gamma \Rightarrow \Delta, \beta}{d:\ \Gamma \Rightarrow \Delta, \alpha \& \beta}$$

is pruned into:

$$\dfrac{e:\ \Gamma' \Rightarrow \Delta'}{d':\ \Gamma \Rightarrow \Delta, \alpha \& \beta}\,weak$$

where $\Gamma' \subseteq \Gamma,\ \Delta' \subseteq \Delta$.

If none of the side formulas of L is introduced by weakening, we apply IH. For example:

$$\dfrac{\dfrac{\Gamma' \Rightarrow \Delta', \alpha}{\Gamma \Rightarrow \Delta, \alpha}\,weak \qquad \dfrac{\Gamma'' \Rightarrow \Delta'', \beta}{\Gamma \Rightarrow \Delta, \beta}\,weak}{d:\ \Gamma \Rightarrow \Delta, \alpha \& \beta}$$

is pruned into:

$$\dfrac{\dfrac{\dfrac{\Gamma' \Rightarrow \Delta', \alpha}{\Gamma^* \Rightarrow \Delta^*, \alpha}\,weak \qquad \dfrac{\Gamma'' \Rightarrow \Delta'', \beta}{\Gamma^* \Rightarrow \Delta^*, \beta}}{\Gamma^* \Rightarrow \Delta^*, \alpha \& \beta}}{d':\ \Gamma \Rightarrow \Delta, \alpha \& \beta}\,weak$$

Here $\Gamma^* \equiv \Gamma', \Gamma'', \quad \Delta^* \equiv \Delta'\Delta''$.

Case 2. L is a one-premise rule. If all side formulas of L are introduced by weakening, then L is deleted. Otherwise weakenings of parametric formulas are moved down; for example:

$$\dfrac{\dfrac{\Gamma' \Rightarrow \beta}{\alpha, \Gamma \Rightarrow \beta}}{\Gamma \Rightarrow \alpha \to \beta}$$

is pruned into:

$$\dfrac{\dfrac{\dfrac{\Gamma' \Rightarrow \beta}{\alpha, \Gamma' \Rightarrow \beta}}{\Gamma' \Rightarrow \alpha \to \beta}}{\Gamma \Rightarrow \alpha \to \beta}$$

⊣

In addition to derivations in LJpm, we consider below *deductions* in LJpm, that is, trees constructed according to the rules of LJpm, with leaves containing arbitrary sequents (not necessarily axioms $\alpha \Rightarrow \alpha$ or $\bot \Rightarrow \alpha$).

LEMMA 11.3. *If $d : \Gamma \Rightarrow \Delta$ is a pruned deduction in LJpm from arbitrary 1-sequents containing no $\vee \Rightarrow$-inferences, then $\Gamma \Rightarrow \Delta$ is an 1-sequent up to weakenings, that is, d ends in:*

$$\frac{\Gamma \Rightarrow \alpha}{d:\ \Gamma \Rightarrow \Delta}\ weak$$

Proof. In a pruned deduction, all rules except $\vee \Rightarrow$ preserve the property of being a 1-sequent; for example:

$$\frac{\dfrac{(\phi \rightarrow \psi)^0, \Gamma' \Rightarrow \phi^0, \Delta'}{(\phi \rightarrow \psi)^0, \Gamma \Rightarrow \phi, \Delta} \quad \dfrac{\psi', \Gamma'' \Rightarrow \Delta''}{\psi, \Gamma \Rightarrow \Delta}}{\phi \rightarrow \psi, \Gamma \Rightarrow \Delta}$$

is in fact

$$\frac{\dfrac{(\phi \rightarrow \psi)^0, \Gamma' \Rightarrow \phi}{\phi \rightarrow \psi, \Gamma \Rightarrow \phi, \delta} \quad \dfrac{\psi, \Gamma'' \Rightarrow \delta}{\psi, \Gamma \Rightarrow \delta}}{\phi \rightarrow \psi, \Gamma \Rightarrow \delta}$$

Indeed both ϕ and ψ are in fact present, since otherwise the whole rule is pruned. Hence $\Delta'' \equiv \delta$ is a single formula, and Δ' is empty. ⊣

DEFINITION 11.2. *Let $d :\ \Gamma \rightarrow \Delta$ is a derivation in LJpm and let L be an antecedent rule in d, with a principal formula π and conclusion $\pi, \Gamma' \Rightarrow \Delta'$. Then L is* movable *if $\pi, \Gamma' \subseteq \Gamma$, and in the case of $L \equiv \rightarrow\Rightarrow$, the succedent parameter formulas are pruned from the left premise:*

$$\frac{\dfrac{(\phi \rightarrow \psi)^0, \Gamma' \Rightarrow \phi}{\phi \rightarrow \psi, \Gamma' \Rightarrow \phi, \Delta'} \quad \psi, \Gamma' \Rightarrow \Delta'}{\phi \rightarrow \psi, \Gamma' \Rightarrow \Delta'}$$

LEMMA 11.4. *Let $d : \Gamma \Rightarrow \Delta$ be a pruned deduction in LJpm from 1-sequents containing an $\vee \Rightarrow$-rule but no non-invertible rules. Then d contains a* movable *rule.*

Proof. Consider the lowermost non-structural rule L in d. If L is an antecedent rule other than $\rightarrow\Rightarrow$, then it is movable. If L is a succedent rule, use IH. Otherwise d ends in:

$$\frac{e:\ \phi \rightarrow \psi, \Gamma' \Rightarrow \Delta', \phi \quad \psi, \Gamma'' \Rightarrow \Delta''}{d:\ \phi \rightarrow \psi, \Gamma \Rightarrow \Delta}\ L$$

If e does not contain $\vee \Rightarrow$, then by Lemma 11.3. the succedent Δ', ϕ contains (up to weakenings) just one formula. Since d is pruned, it should be ϕ; hence L is movable.

If e contains $\vee \Rightarrow$, use IH. ⊣

THEOREM 11.1. *Every derivation of a 1-sequent $\Gamma \Rightarrow \alpha$ in LJpm can be transformed by permutation of movable rules and pruning into a derivation in LJp.*

Proof. We use induction on the number of logical inferences in the given derivation. If all premises of the last rule are 1-sequents (up to weakenings), then apply IH. Otherwise delete from the given derivation all sequents above the lowermost 1-sequent (up to pruning) in every branch. We are left with a deduction d from 1-sequents by invertible rules, since conclusions of non-invertible rules are 1-sequents. If this deduction does not contain $\vee\Rightarrow$-inferences, use Lemma 11.3.; otherwise there is a movable rule. Moving it down and applying IH to its premises concludes the proof. Let us consider possible cases in detail.

Case 1. There is a movable $\&\Rightarrow$-inference. Then d takes the form shown on the left below. Permute it down as shown on the right, to appy IH to d'.

$$
\begin{array}{c}
|\\
\dfrac{\alpha,\Sigma' \Rightarrow \Delta}{\alpha\&\beta,\Sigma' \Rightarrow \Delta}\,L\\
|\\
\alpha\&\beta,\Sigma \Rightarrow \gamma
\end{array}
\qquad
\begin{array}{c}
|\\
\alpha,\Sigma' \Rightarrow \Delta\\
|\\
\dfrac{\dfrac{d' :\ \alpha,(\alpha\&\beta)^0,\Sigma \Rightarrow \gamma}{\alpha\&\beta,(\alpha\&\beta)^0,\Sigma \Rightarrow \gamma}}{\alpha\&\beta,\Sigma \Rightarrow \gamma}\,contr
\end{array}
$$

where the contraction rule and $(\alpha\&\beta)^0$ are present only if the lowermost $\alpha\&\beta$ has other predecessors than the principal formula of L.

Case 2. There is a movable $\vee\Rightarrow$-inference (this case is similar to the previous case):

$$
\begin{array}{c}
|\qquad\qquad |\\
\dfrac{\alpha,\Sigma' \Rightarrow \Delta \quad \beta,\Sigma' \Rightarrow \Delta}{\alpha\vee\beta,\Sigma' \Rightarrow \Delta}\\
|\\
\alpha\vee\beta,\Sigma \Rightarrow \gamma
\end{array}
\qquad
\begin{array}{c}
\begin{array}{cc}
| & |\\
\alpha,\Sigma' \Rightarrow \Delta & \beta,\Sigma' \Rightarrow \Delta\\
| & |
\end{array}\\
\dfrac{\dfrac{\alpha,(\alpha\vee\beta)^0,\Sigma \Rightarrow \gamma \quad \beta,(\alpha\vee\beta)^0,\Sigma \Rightarrow \gamma}{\alpha\vee\beta,(\alpha\vee\beta)^0,\Sigma \Rightarrow \gamma}}{\alpha\vee\beta,\Sigma \Rightarrow \gamma}
\end{array}
$$

Case 3. There is a movable $\rightarrow\Rightarrow$-inference L:

$$
\begin{array}{c}
\dfrac{\dfrac{d' : (\alpha\rightarrow\beta)^0,\Sigma' \Rightarrow \alpha}{(\alpha\rightarrow\beta)^0,\Sigma' \Rightarrow \Delta,\alpha} \quad \beta,\Sigma' \Rightarrow \Delta}{\alpha\rightarrow\beta,\Sigma' \Rightarrow \Delta}\,L\\
|\\
\alpha\rightarrow\beta,\Sigma \Rightarrow \gamma
\end{array}
$$

Move it down:

$$
\dfrac{\dfrac{d' :\ (\alpha\rightarrow\beta)^0,\Sigma' \Rightarrow \alpha}{\alpha\rightarrow\beta,\Sigma \Rightarrow \alpha,\gamma} \quad \begin{array}{c}\beta,\Sigma' \Rightarrow \Delta\\ |\\ \beta,(\alpha\rightarrow\beta)^0,\Sigma \Rightarrow \gamma\end{array}}{\alpha\rightarrow\beta,\Sigma \Rightarrow \gamma}
$$

This concludes the proof. ⊣

Chapter 12

Interpolation Theorem

Now we can prove a proposition having many applications. If E is a formula or sequent, then L_E stands for the list of propositional variables occurring in E plus $\{\top, \bot\}$.

A *Craig interpolant* for an implication $\beta \rightarrow \alpha$ is a formula ι such that:

$$\vdash \beta \rightarrow \iota, \qquad \vdash \iota \rightarrow \alpha \qquad \text{and } L_\iota \subseteq L_\alpha \cap L_\beta.$$

A *Craig interpolant* for a *partition* $\Gamma; \Delta \Rightarrow \alpha$ of a sequent $\Gamma, \Delta \Rightarrow \alpha$ is a formula ι such that:

$$\vdash \Gamma \Rightarrow \iota, \qquad \vdash \iota, \Delta \Rightarrow \alpha \qquad \text{and } L_\iota \subseteq L_\Gamma \cap L_{\Delta \Rightarrow \alpha}.$$

THEOREM 12.1. *(interpolation theorem) (a) If $\Gamma, \Delta \Rightarrow \alpha$ is derivable, then there is a Craig interpolant for $\Gamma; \Delta \Rightarrow \alpha$.*

(b) If $\alpha \rightarrow \beta$ is derivable, then there is a Craig interpolant for $\alpha \rightarrow \beta$.

Proof. Part (b) follows immediately from Part (a). Let us prove Part (a) by induction on a given derivation $d: \ \Gamma, \Delta \Rightarrow \alpha$ in LJp.

Case 0. The d is an axiom $\alpha \Rightarrow \alpha$ or $\bot \rightarrow \alpha$. The interpolant depends on a given partition Γ, Δ of the antecedent. If Γ is empty, define $\iota := \top$, and we have $\Rightarrow \top$ and $\top, \alpha \Rightarrow \alpha$, as well as $\top, \bot \Rightarrow \alpha$. If Γ is non-empty, that is, the whole of the antecedent, define $\iota := \Gamma$ (that is, $\iota := \alpha$ or $\iota := \bot$), and we have $\Gamma \Rightarrow \Gamma$ and $\Gamma \Rightarrow \alpha$.

Case 1. The d ends in an one-premise succedent rule L (that is, $L \equiv \Rightarrow \vee$ or $L \equiv \Rightarrow \rightarrow$). By IH, there is an interpolant ι for the premise, and it is just preserved for the conclusion. If for example $L \equiv \rightarrow \vee$, then we have:

$$\frac{\Gamma, \Delta \Rightarrow \alpha_i}{\Gamma, \Delta \Rightarrow \alpha_0 \vee \alpha_1} \qquad \Gamma \Rightarrow \iota \quad \begin{array}{ll} \iota, \Delta \Rightarrow \alpha_i & \textit{by IH} \\ \iota, \Delta \Rightarrow \alpha_0 \vee \alpha_1 & \textit{by} \Rightarrow \vee \end{array}$$

Case 2. The d ends in a two-premise succedent rule L (that is, $L \equiv \Rightarrow \&$). Take conjunction of interpolants for the premises:

$$\frac{\Gamma, \Delta \Rightarrow \phi \quad \Gamma, \Delta \Rightarrow \psi}{\Gamma, \Delta \Rightarrow \phi \& \psi} \quad \begin{array}{ll} \Gamma \Rightarrow \iota \ \ \Gamma \rightarrow \kappa & \iota, \Delta \Rightarrow \phi \ \ \kappa, \Delta \Rightarrow \psi \qquad \textit{by IH} \\ \Gamma \Rightarrow \iota \& \kappa & \iota \& \kappa, \Delta \Rightarrow \phi \& \psi \quad \textit{by} \Rightarrow \&, \ \& \Rightarrow \end{array}$$

Case 3. The d ends in an antecedent rule L, and L has one premise or the principal formula of L is in Δ. The construction of ι is the same as in Cases 1 and 2: ι is transferred from the premise if L has one premise, and the conjunction of interpolants is taken if L has two premises; for example:

$$\frac{\Gamma, \Delta \Rightarrow \phi \quad \Gamma, \psi, \Delta \Rightarrow \alpha}{\Gamma, \phi \to \psi, \Delta \Rightarrow \alpha} \qquad \frac{\Gamma \Rightarrow \iota \quad \Gamma \to \kappa}{\Gamma \Rightarrow \iota \& \kappa} \qquad \frac{\dfrac{\iota, \Delta \Rightarrow \phi}{\iota \& \kappa, \Delta \Rightarrow \phi} \quad \dfrac{\kappa, \psi, \Delta \Rightarrow \alpha}{\iota \& \kappa, \psi, \Delta \Rightarrow \alpha}}{\iota \& \kappa, \phi \to \psi, \Delta \Rightarrow \alpha}$$

by IH, $\to\Rightarrow, \& \Rightarrow$.

Case 4. The d ends in a two-premise antecedent rule L, and the principal formula of L is in Γ. Then the interpolant has the same principal connective as L. If $L \equiv \vee \Rightarrow$, the argument proceeds as before:

$$\frac{\phi, \Gamma, \Delta \Rightarrow \alpha \quad \psi, \Gamma, \Delta \Rightarrow \alpha}{\phi \vee \psi, \Gamma, \Delta \Rightarrow \alpha} \qquad \frac{\dfrac{\phi, \Gamma \Rightarrow \iota}{\phi, \Gamma \Rightarrow \iota \vee \kappa} \quad \dfrac{\psi, \Gamma \Rightarrow \iota}{\psi, \Gamma \Rightarrow \iota \vee \kappa}}{\phi \vee \psi, \Gamma \Rightarrow \iota \vee \kappa} \qquad \frac{\iota, \Delta \Rightarrow \alpha \quad \kappa, \Delta \Rightarrow \alpha}{\iota \vee \kappa, \Delta \Rightarrow \alpha}$$

If $L \equiv \to\Rightarrow$ with the conclusion partitioned as $\phi \to \psi, \Gamma; \Delta \Rightarrow \alpha$, apply IH to partitions:

$$\Delta; \Gamma \Rightarrow \phi \text{ and } \psi, \Gamma; \Delta \Rightarrow \alpha \ :$$

$$\frac{\Gamma, \Delta \Rightarrow \phi \quad \psi, \Gamma, \Delta \Rightarrow \alpha}{\phi \to \psi, \Gamma, \Delta \Rightarrow \alpha} L \qquad \frac{\dfrac{\iota, \Gamma \Rightarrow \phi \quad \psi, \Gamma \Rightarrow \kappa}{\phi \to \psi, \Gamma, \iota \Rightarrow \kappa}}{\phi \to \psi, \Gamma \Rightarrow \iota \to \kappa} \qquad \frac{\Delta \Rightarrow \iota \quad \kappa, \Delta \Rightarrow \alpha}{\iota \to \kappa, \Delta \Rightarrow \alpha}$$

This concludes the proof. ⊣

12.1. Beth Definability Theorem

One of the applications of the Interpolation Theorem is a criterion of definability of a proposition p satisfying given property α. If there is an *explicit definition* $p \leftrightarrow \iota$, where ι is a formula that does not contain p, then p is obviously unique: $q \leftrightarrow \iota$ implies $p \leftrightarrow q$. Theorem 12.1. below shows that uniqueness implies explicit definability.

DEFINITION 12.1. *A formula α* implicitly defines *a propositional variable p iff the formula:*

$$\alpha \& \alpha[p/q] \to (p \leftrightarrow q) \tag{12.1}$$

is derivable for a new propositional variable q.

Formula (12.1) says that at most one proposition p (up to equivalence) satisfies $\alpha \equiv \alpha[p]$.

THEOREM 12.2. *(Beth definability Theorem). If a formula α implicitly defines p, then α explicitly defines p: there exists a formula ι that does not contain p such that:*

$$\alpha \to (p \leftrightarrow \iota)$$

is derivable.

Proof. Derivability of (12.1) implies derivability of the sequent

$$p, \alpha, \alpha[p/q] \Rightarrow q.$$

Take ι to be an interpolant for the partition $p, \alpha; \alpha[p/q] \Rightarrow q$:

$$p, \alpha \Rightarrow \iota; \qquad \iota, \alpha[p/q] \Rightarrow q$$

This implies (after substituting q by p in the second sequent):

$$\alpha \Rightarrow p \rightarrow \iota \qquad \alpha \Rightarrow \iota \rightarrow p$$

as required. ⊣

Part II

Intuitionistic Predicate Logic

Chapter 13

Natural Deduction System NJ

Predicate logic *terms* are constructed from individuum variables denoted by x, y, z, x_1, y_1, z_1 and so on, by means of function symbols denoted by f, g, h, f_1, g_1, h_1 and so on. Variables and 0-place function symbols (constants) are terms, and if $t_1, \ldots, t_p$, $(p > 0)$ are terms, then $f(t_1, \ldots, t_p)$ is a term for a p-place functional symbol f. *Atomic formulas* are constructed from terms and *predicate symbols* (which are denoted by $P, Q, P_1, Q_1, \ldots$). Propositional letters are atomic formulas (and 0-place predicate symbols); $\perp$ is an atomic formula, and if $t_1, \ldots, t_p$, $p > 0$ are terms, then $P(t_1, \ldots, t_p)$ is an atomic formula for a p-place predicate symbol P. *Formulas* are constructed from atomic formulas by propositional connectives $\&, \vee, \rightarrow$ and quantifiers. Atomic formulas are formulas, and if α, β are formulas, then $\alpha \& \beta, \alpha \vee \beta, \alpha \rightarrow \beta, \forall x \alpha, \exists x \alpha$ are formulas.

A main complexity measure of a formula is its *length*, that is, the number of occurrences of logic connectives; $lth(\alpha) := 0$ for atomic α;

$$lth(\alpha \odot \beta) := lth(\alpha) + lth(\beta) + 1; \qquad lth(\forall x \alpha) \equiv lth(\exists x \alpha) := lth(\alpha) + 1.$$

Quantifiers *bind* occurrences of individuum variables; remaining occurrences are *free*. Let us define the list $FV(\alpha)$ of free variables of the formula α.

If α is atomic, then $FV(\alpha)$ is the list of all variables occurring in α:

$FV(\alpha \odot \beta) := FV(\alpha) \cup FV(\beta)$ for $\odot \equiv \&, \vee, \rightarrow$.

$FV(\forall x \alpha) \equiv FV(\exists x \alpha) := FV(\alpha) - \{x\}$

Recall that the expression $\alpha[x/t]$ stands for the result of substituting a term t for all free occurrences of a variable x in a formula α. For other syntactic expressions $\mathcal{E}$, substitution $\mathcal{E}[x/t]$ is defined componentwise. For example, $(\alpha_1, \ldots \Rightarrow \ldots, \beta_n)[x/t] := \alpha_1[x/t], \ldots \Rightarrow \ldots, \beta_n[x/t]$.

Predicate systems NJ and NK

Add the quantifier rules presented below to propositional versions of these systems and allow $\alpha \equiv \exists x \beta$ in $\bot_c$:

$$\frac{\neg\alpha, \Gamma \Rightarrow \alpha}{\Gamma \Rightarrow \alpha} \bot_c$$

$$NJ := NJp + \{\forall I, \forall E, \exists I, \exists E\}$$

$$NK := NKp + \{\forall I, \forall E, \exists I, \exists E\}$$

$$\frac{\Gamma \Rightarrow \alpha[x/y]}{\Gamma \Rightarrow \forall x \alpha} \forall I \qquad \frac{\Gamma \Rightarrow \forall x \alpha}{\Gamma \Rightarrow \alpha[x/t]} \forall E$$

$$\frac{\Gamma \Rightarrow \alpha[x/t]}{\Gamma \Rightarrow \exists x \alpha} \exists \qquad \frac{\Gamma \Rightarrow \exists x \alpha \quad (\alpha[x/y])^0, \Delta \Rightarrow \theta}{[\Gamma, \Delta] \Rightarrow \theta} \exists E$$

The rules $\forall I$, $\exists E$ have a *proviso for eigenvariable* y: Variable y is not free in the conclusion of the rule and in $\exists x \alpha$.

EXAMPLE 13.1. Standard axioms for quantifiers are obtained by $\forall E, \exists I$:

$$\frac{\forall x \alpha \Rightarrow \forall x \alpha}{\forall x \alpha \Rightarrow \alpha[x/t]} \qquad \frac{\alpha[x/t] \Rightarrow \alpha[x/t]}{\alpha[x/t] \Rightarrow \exists x \alpha}$$

13.1. Derivable Rules

All rules mentioned in Section 2.5. are derivable in NJ. The justification of $\bot_i$ with arbitrary α (Lemma 7.1) is extended by quantifier introduction rules, for example:

$$\frac{\dfrac{\Gamma \Rightarrow \bot}{\Gamma \Rightarrow \alpha[x/y]}}{\Gamma \Rightarrow \forall x \alpha}$$

with a new variable y.

To the rules for analysis into subgoals (Section 2.6.) we add $\forall I, \exists I$, and $\exists \Rightarrow$:

$$\frac{\alpha[x/y], \Gamma \Rightarrow \theta}{\exists x \alpha, \Gamma \Rightarrow \theta} \exists \Rightarrow$$

with a new variable y.

To the rules for direct chaining (Section 2.6.) one adds $\forall E$ and $\exists \Rightarrow$. After this the ADC method for establishing deducibility (Section 2.6.) is described in a natural way. First the goal sequent is analyzed into subgoals, and then direct chaining is applied to each subgoal. The rules $\forall E, \exists I$ are restricted to parameters present in (sub)goals.

All results on heuristics in Section 2.7. extend to predicate logic. In particular NJ is obviously contained in NK.

LEMMA 13.1. *Every sequent derivable in NJ is derivable in NK, and hence it is classically valid.*

⊣

We add $\exists \Rightarrow$ and its inverse to heuristics for $\exists$, as well as a similar relation for $\forall$:

$$\exists x\alpha, \Gamma \Rightarrow \theta \qquad \text{iff } \alpha[x/y], \Gamma \Rightarrow \theta$$

$$\Gamma \Rightarrow \forall x\alpha \qquad \text{iff } \Gamma \Rightarrow \alpha[x/y]$$

with a new variable y.

To heuristics for negation we add:

$$\vdash \neg\exists x\alpha \leftrightarrow \forall x\neg\alpha \tag{13.1}$$

$$\vdash \neg\neg\exists x\alpha \leftrightarrow \neg\forall x\neg\alpha \tag{13.2}$$

Formula (13.1) is proved by ADC, and formula (13.2) immediately follows from it.

The replacement of equivalents (Section 2.8.) has to be modified in a familiar way if it is done in the scope of quantifiers: the formula $(Px \leftrightarrow Qx) \rightarrow (\forall x Px \leftrightarrow \forall x Qx)$ is not even classically valid. Lemma 2.8. is extended as follows.

LEMMA 13.2. *(a)*

$$\vdash \forall x(\alpha \leftrightarrow \beta) \rightarrow (\forall x\alpha \leftrightarrow \forall x\beta)$$

$$\vdash \forall x(\alpha \leftrightarrow \beta) \rightarrow (\exists x\alpha \leftrightarrow \exists x\beta)$$

(b)

$$\vdash \forall x_1 \ldots \forall x_k(\alpha \leftrightarrow \beta) \rightarrow (\gamma[p/\alpha] \leftrightarrow \gamma[p/\beta])$$

where $x_1, \ldots, x_k$ include all free variables of α, β.

Proof. Part (a) is proved by ADC.

Part (b) is proved by induction on γ. Let $Eq := \forall x_1 \ldots \forall x_k(\alpha \leftrightarrow \beta)$. Atomic case $Eq \rightarrow (\alpha \leftrightarrow \beta)$ is proved using $\forall E$. Propositional case $\gamma \equiv \gamma_0 \odot \gamma_1$ is proved using $\forall E$ and Lemma 2.8.(a). The remaining case $\gamma \equiv \mathbf{Q}\delta$ with $\mathbf{Q} \equiv \forall, \exists$ is proved using Part (a):

$$\dfrac{\dfrac{Eq \Rightarrow \delta[p/\alpha] \leftrightarrow \delta[p/\beta]}{Eq \Rightarrow \forall x(\delta[p/\alpha] \leftrightarrow \delta[p/\beta])}\,\forall I}{Eq \Rightarrow \mathbf{Q}x\delta[p/\alpha] \leftrightarrow \mathbf{Q}x\delta[p/\beta]}\,(a)$$

⊣

13.2. Gödel's Negative Translation

Gödel's negative translation still works for predicate logic, but Glivenko's theorem (Theorem 3) does not extend literally: Formula $\forall x\neg\neg Px \rightarrow \neg\neg\forall x Px$ (and intuitionistically equivalent $\neg(\forall x\neg\neg Px \& \neg\forall x Px)$) are derivable in NK, but not in NJ (see Example 14 below). Definition 3 is extended to quantifiers in a natural way: $\forall$ is preserved, $\exists$ is prefixed by $\neg\neg$.

DEFINITION 13.1.

$$\alpha^{neg} := \neg\neg p \qquad \textit{for atomic } \alpha$$

$$(\alpha \& \beta)^{neg} := \alpha^{neg} \& \beta^{neg} \qquad (\alpha \to \beta)^{neg} := \alpha^{neg} \to \beta^{neg}$$

$$(\alpha \vee \beta)^{neg} := \neg(\neg\alpha^{neg} \& \neg\beta^{neg})$$

$$(\forall x \alpha)^{neg} := \forall x(\alpha^{neg}) \qquad (\exists x \alpha)^{neg} := \neg\forall x \neg(\alpha^{neg}).$$

In connection with the last clause, note that by (13.2)

$$\vdash \neg\forall x \neg\alpha \leftrightarrow \neg\neg\exists x \alpha.$$

To justify the negative translation, we use the analog of Lemma 3(a):

$$\vdash \neg\neg\forall x \neg\alpha \leftrightarrow \forall x \neg\alpha \tag{13.3}$$

which is proved using ADC.

DEFINITION 13.2. *A formula is* negative *if it does not contain* $\vee, \exists$, *and all atomic subformulas are negated.*

Note that $\neg\alpha$ is not necessarily negative in this sense.

LEMMA 13.3. *In NJ:* $\vdash \alpha \leftrightarrow \neg\neg\alpha$ *for every negative formula* α.

Proof. Extend the proof of Lemma 3(b) using (13.3). ⊣

LEMMA 13.4. *Every rule of NK is stable under Gödel's negative translation.*

Proof. Extend the proof of Lemma 3 using relations:

$$\alpha[x/t] \vdash \neg\forall x \neg\alpha \tag{13.4}$$

$$\neg\forall x \neg\alpha, \forall y(\alpha \to \theta^{neg}) \vdash \theta^{neg} \tag{13.5}$$

(13.4) is obtained by ADC. To obtain (13.5), prefix the last θ^{neg} by $\neg\neg$ (Lemma 13.2.) and reduce this goal to a subgoal:

$$\forall y(\alpha[x/y] \to \theta^{neg}), \neg\theta^{neg} \vdash \forall x \neg\alpha.$$

This is proved by ADC. ⊣

THEOREM 13.1. *A sequent* $\Gamma \Rightarrow \Delta$ *is derivable in NK iff* $\Gamma^{neg} \Rightarrow \Delta^{neg}$ *is derivable in NJ.*

Proof. Exactly as Theorem 3. ⊣

The next Theorem illustrates Glivenko-style results that can be obtained by Gödel's negative translation.

THEOREM 13.2. *(a) A negative formula is derivable (in NJ) iff it is derivable in NK*

(b) If a formula α does not contain universal quantifiers, then $\vdash \neg\alpha$ iff $\neg\alpha$ is derivable in NK.

Proof. To reduce Part (b) to Part (a), prove by induction on α that $\neg\alpha$ is equivalent to a negative formula if α does not contain universal quantifiers. Then note that Gödel's negative translation does not change negative formulas ν:

$$\vdash \nu^{neg} \leftrightarrow \nu$$

and use Theorem 13.2.. ⊣

13.3. Program Interpretation of NJ

Add the following clauses to Section 4.1.:

$c\mathbf{r}\exists x\alpha$ iff $c = \mathbf{p}(d, a)$, d is an individuum and $a\mathbf{r}\alpha[x/d]$,

$c\mathbf{r}\forall x\alpha$ iff c is a function and for every individuum d, $c(d)\mathbf{r}\alpha[x/d]$.

Examples.

Realization t of $\forall x\exists y\alpha$ is a function (program) that for every x produces value $\mathbf{p}_0(t(x))$ and a realization $\mathbf{p}_1(t(x))$ of the remaining part of the formula $\alpha[y/\mathbf{p}_0(t(x))]$.

Realization t of $\exists x\alpha \rightarrow \exists y\beta$ is a program that for every x satisfying α and every realization z of $\alpha[x]$ produces a value $t(\mathbf{p}(x, z))$ that is a pair consisting of a value $\mathbf{p}_0(t(\mathbf{p}(x, z)))$ satisfying β and a realization $\mathbf{p}_1(t(\mathbf{p}(x, z)))$ of $\beta[y/\mathbf{p}_0(t(\mathbf{p}(x, z)))]$.

EXERCISE 13.1. What are realizations of the following formulas:

$$\exists x\alpha \rightarrow \forall z\exists y\beta; \qquad \forall z(\exists x\alpha \rightarrow \exists y\beta)$$

$$\forall x\phi \rightarrow \phi[x/t],\ \forall x(r \rightarrow \phi) \rightarrow (r \rightarrow \forall x\phi),\ x \text{ is not free in } r$$

$$\phi[x/t] \rightarrow \exists x\phi,\ \forall x(\phi \rightarrow r) \rightarrow (\exists x\phi \rightarrow r),\ x \text{ is not free in } r$$

For the language of deductive terms (see Section 4.2.), we distinguish individuum variables $x, y, \dots$ of predicate logic from typed variables $x^\alpha, y^\alpha, \dots$ of the language of deductive terms. We add a new construction $E_{y,u^{\phi[y/x]}}$ to account for $\exists$-elimination. Assume that $t \equiv \mathbf{p}(t_0, t_1)$ realizes $\exists x\phi$ and s^θ is a realization of a formula θ, depending on parameters $y, u^{\phi[y/x]}$ for possible values of x and possible realizations of $\phi[y/x]$. Then by BHK-interpretation, $s[y/t_0, u/t_1]$ is a realization of θ. This is expressed by the equation:

$$E_{y,u^{\phi[y/x]}}(\mathbf{p}(t_0, t_1), s^\theta) = s[y/t_0, u/t_1]. \tag{13.6}$$

that is added to equations (4.1–4.3) together with the version of equation (4.2) corresponding to $\forall$-elimination:

$$(\lambda x.t^\alpha)(u) = t[x/u] \tag{13.7}$$

where x is an individuum variable and t is a term of predicate logic.

Assignment rules for predicate logic:

$$\frac{\mathbf{z} : \Gamma \Rightarrow t : A[x/y]}{\mathbf{z} : \Gamma \Rightarrow \lambda y.t : \forall x\phi} \forall I \qquad \frac{\mathbf{z} : \Gamma \Rightarrow t : \forall x\phi}{\mathbf{z} : \Gamma \Rightarrow t(t') : \phi[x/t']} \forall E$$

$$\frac{\mathbf{z} : \Gamma \Rightarrow t_1 : \phi[x/t_0]}{\mathbf{z} : \Gamma \Rightarrow \mathbf{p}(t_0, t_1) : \exists x\phi} \exists I$$

$$\frac{\mathbf{z} : \Gamma \Rightarrow t : \exists x\phi \quad (u^{\phi[x/y]} : \phi[x/y])^0, \mathbf{z}' : \Delta \Rightarrow t_1 : \theta}{\mathbf{z}, \mathbf{z}' : [\Gamma, \Delta] \Rightarrow E_{y,u}(t, t_1) : \theta} \exists E$$

with a suitable proviso in $\forall I, \exists E$. This extends the assignment $\mathcal{T}$ of deductive terms (Section 4.2.) to derivations in predicate logic.

EXAMPLE 13.2. Using variables $z^{\exists x\alpha\&\forall x\beta}, u^{\alpha[x/y]}$, we have:

$$\dfrac{\dfrac{\dfrac{z : \exists x\alpha\&\forall x\beta \Rightarrow z : \exists x\alpha\&\forall x\beta}{\exists x\alpha\&\forall x\beta \Rightarrow \mathbf{p}_0(z) : \exists x\alpha} \quad \dfrac{u : \alpha[x/y], z : \exists x\alpha\&\forall x\beta \Rightarrow \mathbf{k} : \alpha[x/y]\&\beta[x/y]}{u : \alpha[x/y], \exists x\alpha\&\forall x\beta \Rightarrow \mathbf{p}(y, \mathbf{k}) : \exists x(\alpha\&\beta)}}{\exists x\alpha\&\forall x\beta \Rightarrow D_{y,u}(\mathbf{p}_0 z, \mathbf{p}(y, \mathbf{k})) : \exists x(\alpha\&\beta)}}{\Rightarrow \lambda z.D_{y,u}(\mathbf{p}_0 z, \mathbf{p}(y, \mathbf{k})) : \exists x\alpha\&\forall x\beta \rightarrow \exists x(\alpha\&\beta)}$$

where $\mathbf{k} \equiv \mathbf{p}(u, (\mathbf{p}_1 z)(y))$ and the right uppermost sequent is derived as follows:

$$\dfrac{u : \alpha[x/y] \Rightarrow u : \alpha[x/y] \quad \dfrac{\dfrac{z : \exists x\alpha\&\forall x\beta \Rightarrow z : \exists x\alpha\&\forall x\beta}{z : \exists x\alpha\&\forall x\beta \Rightarrow \mathbf{p}_1 z : \forall x\beta}}{z : \exists x\alpha\&\forall x\beta \Rightarrow (\mathbf{p}_1 z)(y) : \beta[x/y]}}{u : \alpha[x/y], z : \exists x\alpha\&\forall x\beta \Rightarrow \mathbf{k} : \alpha[x/y]\&\beta[x/y]}$$

The Curry–Howard isomorphism (Section 4.3.) between deductive terms and natural deductions can be extended to NJ. Operation $\mathcal{D}$ (Definition 4.3.) is extended in a natural way to new deductive terms t.

If $t^{\forall x\phi} \equiv \lambda y u^{\phi[x/y]}$, then:

$$\frac{\mathcal{D}(u) : \Gamma \Rightarrow \phi[x/y]}{\mathcal{D}(\lambda y.u) : \Gamma \Rightarrow \forall x\phi} \forall I$$

If $t^{\alpha[x/t']} \equiv s^{\forall x\alpha}(t')$, then:

$$\frac{\mathcal{D}(s) : \Gamma \Rightarrow \forall x\alpha}{\mathcal{D}(s(t')) : \Gamma \Rightarrow \alpha[x/t']} \forall E$$

If $t^{\exists x\alpha} \equiv \mathbf{p}(s^{\alpha[x/t']}, u)$, then:

$$\frac{\mathcal{D}(s) : \Gamma \Rightarrow \alpha[x/t]}{\mathcal{D}(\mathbf{p}(s, t)) : \Gamma \Rightarrow \exists x\alpha} \exists I$$

If $t^{\theta} \equiv E_{y,u^{\alpha[x/y]}}(t_0^{\exists x\alpha}, t_1^{\theta})$, then:

$$\frac{\mathcal{D}(t_0) : \ \Gamma \Rightarrow \exists x\alpha \quad \mathcal{D}(t_1) : \ \alpha[x/y], \Delta \Rightarrow \theta}{\mathcal{D}(E(t_0, t_1)) : \ [\Gamma, \Delta] \Rightarrow \theta}$$

Conversions for deductive terms are relations (4.1–4.3), a conversion for $\forall$ similar to (4.2):

$$(\lambda y.t)(u) = t[y/u] \tag{13.8}$$

and the following conversion for $\exists$:

$$E_{x_0,x_1}(\mathbf{p}(t_0, t_1), t_2) \ conv \ t_2[x_0/t_0, x_1/t_1] \tag{13.9}$$

Corresponding conversions for natural deduction follow.

$\forall$-conversion:

$$\frac{\dfrac{\begin{array}{c} y \quad \vdots \\ e : \ \Gamma \Rightarrow t : \alpha[x/y] \end{array}}{\Gamma \Rightarrow \lambda y.t : \forall x\alpha}}{\Gamma \Rightarrow (\lambda y.t)(t') : \alpha[x/t']} \qquad conv \qquad \begin{array}{c} t' \quad \vdots \\ e[y/t'] : \ \Gamma \Rightarrow t[y/t'] : \alpha[x/t'] \end{array}$$

Free occurrences of y in the deduction e are replaced by t'.

$\exists$-conversion:

$$\frac{\dfrac{\Gamma \Rightarrow t_1 : \alpha[x/t_0]}{\Gamma \Rightarrow \mathbf{p}(t_0, t_1) : \exists x\alpha} \qquad \begin{array}{c} u^{\alpha[x/y]} : \alpha[x/y] \Rightarrow u^{\alpha[x/y]} : \alpha[x/y] \\ y, u \quad \vdots \\ (u^{\alpha[x/y]} : \alpha[x/y])^0, \Delta \Rightarrow s : \theta \end{array}}{[\Gamma, \Delta] \Rightarrow E_{y,u}(\mathbf{p}(t_0, t_1), s) : \theta}$$

$$conv \qquad \begin{array}{c} \Gamma \Rightarrow t_1 : \alpha[x/t_0] \\ t_0, t_1 \quad \vdots \\ [\Gamma^0, \Delta] \Rightarrow s[y/t_0, u/t_1] : \theta \end{array}$$

Free occurrences of y are replaced by t_0, and the assumption $\alpha[x/y]$ is replaced by Γ. The definitions in Section 5.1. (ancestral relations in a natural deduction, substitution of a deduction for an assumption, reductions for deductive terms and deductions, and so on) are extended in a natural way. After this Lemma 5.2. and Theorem 5.2. extend to NJ together with their proofs. The notions connected with conversions, reductions, and normal form are defined exactly as in Section 5.1..

THEOREM 13.3. *(Curry–Howard isomorphism between terms and natural deductions in predicate logic)*

(a) Every natural deduction d in NJ uniquely defines $\mathcal{T}(d)$ and vice versa: Every deductive term t uniquely defines a natural deduction $\mathcal{D}(t)$.

(b) Cuts in d uniquely correspond to redexes in $\mathcal{T}(d)$ and vice versa.

(c) Every conversion in d uniquely corresponds to a conversion in $\mathcal{T}(d)$, and reduction sequences for d uniquely correspond to reduction sequences for $\mathcal{T}(d)$ and vice versa.

(d) A deduction d is normal iff $\mathcal{T}(d)$ is normal and vice versa.

EXAMPLE 13.3. Consider the following deduction of a sequent $\alpha \Rightarrow \bot$, where:

$\alpha := \exists x \forall y(Px \vee \neg Py)$, $\beta := \forall y \neg Py \& \neg \forall y \neg Py$

$$\dfrac{\dfrac{\dfrac{\dfrac{\beta \Rightarrow \beta}{\beta \Rightarrow \neg\forall y\neg Py} \quad \dfrac{\beta \Rightarrow \beta}{\beta \Rightarrow \forall y\neg Py}}{\beta \Rightarrow \bot}}{\Rightarrow \beta \rightarrow \bot} \qquad \dfrac{d_0 : \ \neg\alpha \Rightarrow \neg\forall y\neg Py \quad d_1 : \ \neg\alpha \Rightarrow \forall y\neg Py}{\neg\alpha \Rightarrow \beta}}{d : \ \neg\alpha \Rightarrow \bot}$$

Where d_0, d_1 are as follows:

$$\dfrac{\dfrac{\neg\alpha \Rightarrow \neg\alpha \qquad \dfrac{\dfrac{\dfrac{\dfrac{\forall y\neg Py \Rightarrow \forall y\neg Py}{\forall y\neg Py \Rightarrow \neg Pa}}{\forall y\neg Py \Rightarrow Px \vee \neg Pa}}{\forall y\neg Py \Rightarrow \forall y(Px \vee \neg Py)}}{\forall y\neg Py \Rightarrow \alpha}}{\neg\alpha, \forall y\neg Py \Rightarrow \bot}}{d_0 : \ \neg\alpha \Rightarrow \neg\forall y\neg Py}$$

$$\dfrac{\dfrac{\dfrac{\neg\alpha \Rightarrow \neg\alpha \qquad \dfrac{\dfrac{\dfrac{Pb \Rightarrow Pb}{Pb \Rightarrow Pb \vee \neg Py}}{Pb \Rightarrow \forall y(Pb \vee \neg Py)}}{Pb \Rightarrow \exists x\forall y(Px \vee \neg Py)}}{\neg\alpha, Pb \Rightarrow \bot}}{\neg\alpha \Rightarrow \neg Pb}}{d_1 : \ \neg\alpha \Rightarrow \forall y\neg Py}$$

Deduction d ends in a cut, and it is converted into the following deduction d_2:

$$\dfrac{\dfrac{\dfrac{d_0 : \neg\alpha \Rightarrow \neg\forall y\neg Py \quad d_1 : \neg\alpha \Rightarrow \forall y\neg Py}{\neg\alpha \Rightarrow \neg\forall y\neg Py \& \forall y\neg Py}}{\neg\alpha \Rightarrow \neg\forall y\neg Py} \qquad \dfrac{\dfrac{\neg\alpha \Rightarrow \neg\forall y\neg Py \quad \neg\alpha \Rightarrow \forall y\neg Py}{\neg\alpha \Rightarrow \neg\forall y\neg Py \& \forall y\neg Py}}{\neg\alpha \Rightarrow \forall y\neg Py}}{d_2 : \ \neg\alpha \Rightarrow \bot}$$

Converting two &-cuts in this deduction, we obtain d_3:

$$\dfrac{\dfrac{\dfrac{Ax\ \neg\alpha \qquad \dfrac{\dfrac{\dfrac{\dfrac{\forall y\neg Py \Rightarrow \forall y\neg Py}{\forall y\neg Py \Rightarrow \neg Pa}}{\forall y\neg Py \Rightarrow Px \vee \neg Pa}}{\forall y\neg Py \Rightarrow \forall y(Px \vee \neg Py)}}{\forall y\neg Py \Rightarrow \alpha}}{\neg\alpha, \forall y\neg Py \Rightarrow \bot}}{d_0 :\ \neg\alpha \Rightarrow \neg\forall y\neg Py} \qquad \dfrac{\dfrac{\dfrac{Ax\ \neg\alpha \qquad \dfrac{\dfrac{\dfrac{Pb \Rightarrow Pb}{Pb \Rightarrow Pb \vee \neg Py}}{Pb \Rightarrow \forall y(Pb \vee \neg Py)}}{Pb \Rightarrow \exists x\forall y(Px \vee \neg Py)}}{\neg\alpha, Pb \Rightarrow \bot}}{\neg\alpha \Rightarrow \neg Pb}}{d_1 :\ \neg\alpha \Rightarrow \forall y\neg Py}}{d_3 :\ \neg\alpha \Rightarrow \bot}$$

Convert $\rightarrow$-cut at the end of this deduction:

$$\dfrac{\neg\alpha \Rightarrow \neg\alpha \qquad \dfrac{\dfrac{\dfrac{\dfrac{\dfrac{\dfrac{\dfrac{\neg\alpha \Rightarrow \neg\alpha \qquad \dfrac{\dfrac{\dfrac{Pb \Rightarrow Pb}{Pb \Rightarrow Pb \vee \neg Py}}{Pb \Rightarrow \forall y(Pb \vee \neg Py)}}{Pb \Rightarrow \exists x\forall y(Px \vee \neg Py)}}{\neg\alpha, Pb \Rightarrow \bot}}{\neg\alpha \Rightarrow \neg Pb}}{d_1 :\ \neg\alpha \Rightarrow \forall y\neg Py}}{\neg\alpha \Rightarrow \neg Pa}}{\neg\alpha \Rightarrow Px \vee \neg Pa}}{\neg\alpha \Rightarrow \forall y(Px \vee \neg Py)}}{\neg\alpha \Rightarrow \alpha}}{d_4 :\ \neg\alpha \Rightarrow \bot}$$

Conversion of $\forall y\neg Py$-cut results in the following normal form:

$$\dfrac{\neg\alpha \Rightarrow \neg\alpha \qquad \dfrac{\dfrac{\dfrac{\dfrac{\dfrac{\neg\alpha \Rightarrow \neg\alpha \qquad \dfrac{\dfrac{\dfrac{Pa \Rightarrow Pa}{Pa \Rightarrow Pa \vee \neg Py}}{Pa \Rightarrow \forall y(Pa \vee \neg Py)}}{Pa \Rightarrow \exists x\forall y(Px \vee \neg Py)}}{\neg\alpha, Pa \Rightarrow \bot}}{\neg\alpha \Rightarrow \neg Pa}}{\neg\alpha \Rightarrow Px \vee \neg Pa}}{\neg\alpha \Rightarrow \forall y(Px \vee \neg Py)}}{\neg\alpha \Rightarrow \alpha}}{d_5 :\ \neg\alpha \Rightarrow \bot}$$

Let us repeat this for deductive terms. Using variables $z^{\neg\alpha}, u^{\forall y\neg Py}, v^{Pb}, w^{\beta}$, we obtain

$$\mathcal{T}(d_0) \equiv \lambda u.z(\mathbf{p}(x, \lambda a.\mathbf{k}_1(u(a)))); \qquad \mathcal{T}(d_1) \equiv \lambda b.\lambda v.z(\mathbf{p}(b, \lambda y.\mathbf{k}_0 v))$$

$$\mathcal{T}(d) \equiv (\lambda w.(\mathbf{p}_1 w)(\mathbf{p}_0 w))(\mathbf{p}(\mathcal{T}(d_0), \mathcal{T}(d_1))).$$

Then:

$$\mathcal{T}(d)\ \mathit{conv}\ (\mathbf{p}_1(\mathbf{p}(\mathcal{T}(d_0), \mathcal{T}(d_1))))(\mathbf{p}_0(\mathbf{p}(\mathcal{T}(d_0), \mathcal{T}(d_1)))) \equiv \mathcal{T}(d_2)\ \mathit{red}$$

$$(\mathcal{T}(d_0))(\mathcal{T}(d_1)) \equiv \mathcal{T}(d_3) \; conv$$
$$z(\mathbf{p}(x, \lambda y.\mathbf{k}_0((\mathcal{T}(d_1))(a)))) \equiv \mathcal{T}(d_4) \equiv$$
$$z(\mathbf{p}(x, \lambda y.\mathbf{k}_0((\lambda b.\lambda v.z(\mathbf{p}(b, \lambda y.\mathbf{k}_0 v)))(a))))$$
$$red \; z(\mathbf{p}(x, \lambda y.\mathbf{k}_0(\lambda v.z(\mathbf{p}(a, \lambda y.\mathbf{k}_0 v))))) \equiv \mathcal{T}(d_5) \equiv |d|$$

To extend the Normalization theorem (Theorem 5.3.) to NJ, let us preserve the definition of *cutrank* of a redex as the length of its cut formula:

$$cutrank((\lambda x.t^{\alpha[x/y]})(t')) := lth(\forall x\alpha) = lth(\alpha) + 1;$$

$$cutrank(E_{y,u}(\mathbf{p}(t_0, t_1^{\alpha[x/t]}), s)) = lth(\exists x\alpha) = lth(\alpha) + 1$$

Lemma 5.3. and Theorem 5.3. carry over to NJ together with proofs. We must add one more kind of a "new" redex to Lemma 5.3.(a):

$$t \equiv E(x, s) \text{ and } s \equiv \mathbf{p}(u_0, u_1)$$

THEOREM 13.4. *(normalization theorem). (a) Every deductive term t can be normalized.*

(b) Every natural deduction d in NJ can be normalized.

Basic properties of normal deductions (Theorem 5.4.) are preserved, too, and hence $\vee$-property and Harrop's theorem are extended to NJ together with their proofs. In addition we have the $\exists$-property.

THEOREM 13.5. *(Properties of normal deductions). Let* $d : \Gamma \Rightarrow \gamma$ *be a normal deduction in NJ.*

(a) If d ends in an elimination rule, then the main branch contains only elimination rules, it begins with an axiom, and every sequent in it is of the form $\Gamma' \Rightarrow \alpha$*, where* $\Gamma' \subset \Gamma$ *and* α *is some formula.*

(a1) In particular the axiom at the top of the main branch is of the form $\alpha \Rightarrow \alpha$ *where* $\alpha \in \Gamma$.

(b) If $\Gamma = \emptyset$*, then d ends in an introduction rule.*

THEOREM 13.6. *(disjunction property, Harrop's theorem). (a) If* $\vdash \alpha_0 \vee \alpha_1$*, then* $\vdash \alpha_i$ *for some* $i = 0, 1$.

(b) If $\vdash \neg\beta \Rightarrow \alpha_0 \vee \alpha_1$*, then* $\neg\beta \Rightarrow \vdash \alpha_i$ *for some* $i = 0, 1$.

(c) If $\vdash \exists x\alpha$*, then* $\vdash \alpha[x/t]$ *for some t.*

(d) If $\vdash \neg\beta \Rightarrow \exists x\alpha$*, then* $\neg\beta \Rightarrow \vdash \alpha[x/t]$ *for some t.*

Chapter 14

Kripke Models for Predicate Logic

Definitions and results of Chapter 7 are extended here to predicate logic.

DEFINITION 14.1. *A* Kripke frame *for intuitionistic predicate logic is a triple:*

$$< W, R, D >$$

where $< R, W >$ *is a propositional intuitionistic Kripke frame (Definition 7) and* D *is a* domain function *assigning to every* $w \in W$ *a* non-empty *set* $D(w)$ *expanding with respect to* R*:*

$$Rww' \text{ implies } D(w) \subseteq D(w')$$

To every world w of a Kripke frame $< R, W, D >$ corresponds an extension L_w of the language of predicate logic obtained by adding constants for all elements of the domain $D(w)$. We identify such a constant with the corresponding element of $D(w)$. Hence *sentences* of L_w are formulas containing no free occurrences of variables but possibly containing objects of $D(w)$.

DEFINITION 14.2. *A* Kripke model *for intuitionistic predicate logic is a 4-tuple* $< W, R, D, V >$, *where* $< R, W, D >$ *is a Kripke frame and* V *assigns a function* $V(f)$ *to every function symbol* f *and a predicate* $V(P)$ *to every predicate symbol* P, *so that the following conditions are satisfied.*

For all $w \in W$, $d_1, \ldots, d_n \in D(w)$ (and n-ary f, P):

$$V(f)(d_1, \ldots, d_n, w) \in D(w) \quad \text{and} \quad V(P)(d_1, \ldots, d_n, w) \in \{0, 1\}$$

$$Rww' \text{ implies } V(f)(d_1, \ldots, d_n, w') = V(f)(d_1, \ldots, d_n, w) \tag{14.1}$$

$$Rww' \text{ and } V(P)(d_1, \ldots, d_n, w) = 1 \text{ implies } V(P)(d_1, \ldots, d_n, w') = 1 \tag{14.2}$$

In other words:

$$f_w = \lambda d_1 \ldots \lambda d_n . V(f)(d_1, \ldots, d_n, w)$$

is an n-ary function on $D(w)$ and:

$$P_w = \lambda d_1 \ldots \lambda d_n . V(P)(d_1, \ldots, d_n, w)$$

is an n-ary predicate on $D(w)$. Moreover f_w, P_w are monotone in w: The value true for P and all values of f are preserved in the move to an accessible world. The values of variable-free terms and sentences in a Kripke model are defined by the following recursion.

DEFINITION 14.3. *(a) The value $V(t, w) \in D(w)$ of a constant term $t \in D(w)$:*
Constants: $V(d, w) := d$ *if* $d \in D(w)$
Composite terms: $V(f(t_1, \ldots, t_p), w) := V(f)(V(t_1, w), \ldots, V(t_p, w), w)$
(b) The value $V(\alpha, w) \in \{0, 1\}$ for a sentence $\alpha \in L_w$. First atomic formulas:

$$V(P(t_1, \ldots, t_n), w) := V(P)(V(t_1, w), \ldots, V(t_n, w), w)$$

$\&, \vee, \rightarrow, \perp$: *the same clauses as in the Definition 7.*

$\exists$: $V(\exists x\phi) = 1$ *iff* $V(\phi[x/d], w) = 1$ *for some* $d \in D(w)$

$\forall$: $V(\forall x\phi) = 1$ *iff for all* w' *with* Rww' *we have* $V(\phi[x/d], w') = 1$ *for all* $d \in D(w')$.

A formula α is valid *if for every Kripke model* $< W, R, D, V >$, *every world* $w \in W$, *and every substitution* $[x_1/d_1, \ldots, x_n/d_n]$ *of objects in* $D(w)$ *for free variables of* α, *we have* $V(\alpha[x_1/d_1, \ldots, x_n/d_n], w) = 1$. *A sequent* $\Gamma \Rightarrow \Delta$ *is* valid *iff a formula* $\&\Gamma \rightarrow \vee\Delta$ *is valid.*

EXAMPLE 14.1. Let us refute the formula $\forall x \neg\neg Px \rightarrow \neg\neg\forall x P(x)$.

Consider the following Kripke model $\langle W, R, D, V\rangle$ with a constant individuum domain

$D(w)$= the set of positive natural numbers, $W = \omega$ and:

$$V(P)(d, i) = 1 \text{ iff } d < i; \qquad Rij \text{ iff } i \leq j$$

Then $V(\forall x Px, i) = 0$ for every i, since $V(P(i+1), i) = 0$. Hence $V(\neg\forall x Px, 0) = 1$. Moreover $V(\neg\neg Pd, i) = 1$ for all d, i, since $V(Pd, i) = 1$ for $i > d$. Hence $V(\forall x\neg\neg Px, 0) = 1$ and:

$$V(\forall x \neg\neg Px \rightarrow \neg\neg\forall x P(x), 0) = 0$$

Monotonicity (14.1, 14.2), extends to all terms and formulas.

LEMMA 14.1. *Assume* Rww'. *Then:*

$$V(t, w') = V(t, w) \text{ and } V(\alpha, w) = 1 \text{ implies } V(\alpha, w') = 1$$

Proof. To extend the inductive proof of Lemma 7, assume Rww'.

If $V(\exists x\phi, w) = 1$, then the witness $d \in D(w)$ satisfying $\phi[x/d]$ at w is good for w', since $d \in D(w) \subseteq D(w')$.

If $V(\forall x\phi, w) = 1$, take arbitrary w'' with $Rw'w''$ and $d \in D(w')$. Then Rww'' (transitivity), $d \in D(w'')$ (monotonicity of D), and $V(\phi[x/d], w'') = 1$ by definition of $V(\forall x\phi, w) = 1$. Hence $V(\forall x\phi, w') = 1$ as required. ⊣

LEMMA 14.2. *System NJ is sound: All rules preserve validity; hence every derivable sequent is valid.*

Proof. In view of Theorem 7.1., it is enough to check quantifier rules. Consider arbitrary Kripke model $< W, R, D, V >$, take arbitrary $w \in W$ and arbitrary substitution $\sigma \equiv [x_1/d_1, \ldots, x_n/d_n]$ of objects in $D(w)$ for free variables of the conclusion of the rule considered.

Case 1. $\exists I$:

$$\frac{\Gamma \Rightarrow \alpha[x/t]}{\Gamma \Rightarrow \exists x\alpha}$$

Extend given substitution σ to σ^+ by arbitrary substitution of an element of $D(w)$ for all variables $y \in FV(t)$ that are not free in the conclusion $\Gamma \Rightarrow \exists x\alpha$. We have $V((\Gamma \Rightarrow \alpha[x/t])\sigma^+, w) = 1$; hence $V(\Gamma\sigma \Rightarrow (\alpha[x/V(t\sigma^+, w)])\sigma, w) = 1$, and $V(t\sigma^+, w)$ is a witness for $\exists x$ in $V(\Gamma\sigma \Rightarrow (\exists x\alpha)\sigma, w) = 1$.

Case 2. $\forall I$:

$$\frac{\Gamma \Rightarrow \alpha[x/y]}{\Gamma \Rightarrow \forall x\alpha}$$

To prove $V((\&\Gamma \rightarrow \forall x\alpha)\sigma, w) = 1$, assume that Rww' (for $\rightarrow$) and $Rw'w''$ (for $\forall x$), $V(\&\Gamma\sigma, w') = 1$ and take arbitrary $d \in D(w'')$. Take $\sigma^+ := \sigma \cup [y/d]$. Then $V((\&\Gamma \rightarrow \alpha[x/y])\sigma^+, w'') = 1$ by IH; hence $V(\alpha[x/y]\sigma^+, w'') = 1$, so $V(\alpha\sigma[x/d], w) = 1$ as required.

Cases of $\forall E, \exists E$ are treated similarly. ⊣

14.1. Pointed Models, Frame Conditions

Material in Section 7.2. applies to predicate logic without essential changes. In particular, pointed frames and pointed models are defined exactly as before.

DEFINITION 14.4. *A* pointed frame *is an ordered 4-tuple $< G, W, R, D >$, where $< W, R, D >$ is a frame, $G \in W$ and RGw for all $w \in W$. A* pointed model *is a tuple $M =< G, W, R, D, V >$ where $< G, W, R, D >$ is a pointed frame and V is a valuation on $< W, R, D >$. Truth in M is truth in the world G:*

$$M \models \varphi \textit{ iff } V(G, \varphi) = 1$$

LEMMA 14.3. *A formula is valid iff it is true in all pointed models, iff it is true in all pointed models where accessibility relation is a partial order.*

Proof. The implication in one direction is obvious. For the other direction, combine proofs of Lemma 7.2. and Theorem 7.2. ⊣

Let us illustrate the use of frame conditions.

DEFINITION 14.5. *A Kripke frame* $< W, R, D >$ has constant domains *if*

$$D(w) = D(w') \text{ for all } w, w' \in W \tag{14.3}$$

LEMMA 14.4. *A pointed frame has constant domains iff every formula of the form*

$$\forall x(\gamma \vee \alpha) \to \gamma \vee \forall x\alpha \tag{14.4}$$

with $x \notin FV(\gamma)$ *is valid on this frame.*

Proof. Assume first that an arbitrary Kripke frame $< W, R, D >$ has constant domains, and take arbitrary valuation V for this frame and $w \in W$. Assume that (14.4) is a sentence to simplify notation. To prove $V((14.4), w) = 1$, take w' with Rww' and assume $V(\forall x(\gamma \vee \alpha), w') = 1$. If $V(\gamma, w') = 1$, we are done. Otherwise assume (for $\forall x\alpha$) that $w'' \in W$ and $Rw'w''$. Take arbitrary $d \in D(w'')$. By (14.3), $d \in D(w')$; hence $V(\gamma \vee \alpha[x/d], w') = 1$. With $V(\gamma, w') = 0$ this implies $V(\alpha[x/d], w') = 1$, and $V(\alpha[x/d], w'') = 1$ holds by monotonicity, so (14.4) is valid.

Second consider a pointed frame $< G, W, R, D >$, a world $w_0 \in W$, and an object $d_0 \in D(w_0) - D(G)$. For every $w \in W$ take

$$V(p, w) := 1 \text{ iff } d_0 \in D(w); \qquad V(Qd, w) := 1 \text{ iff } d \in D(w) - \{d_0\}$$

and prove that (14.4) is refuted for $\gamma := p, \alpha := Qx$. First for all w, if $d_0 \in D(w)$, then $V(p, w) = 1$ and $V(p \vee Qd, w) = 1$. If $d_0 \notin D(w)$, then $V(Qd, w) = 1$ for all $d \in D(w)$ and again $V(p \vee Qd, w) = 1$. Hence $V(\forall x(p \vee Qx), G) = 1$.

On the other hand, $V(p, G) = 0$, since $d_0 \notin D(G)$, and $V(\forall xQx, G) = 0$, since $V(Qd_0, w_0) = 0$. Hence $V(p \vee \forall xQx, G) = 0$. ⊣

In fact NJ+(14.4) is sound and complete for Kripke frames with constant domains , but we shall not prove this here.

Chapter 15

Systems LJm, LJ

The LJm is the predicate version of LJpm. It has the same formulas as NJ, but sequents $\alpha_1, \ldots, \alpha_m \Rightarrow \beta_1, \ldots, \beta_n$ $(m \geq 0, n > 0)$ may have multiple succedents.

The axioms and propositional rules are the same as in LJp.

Four quantifier rules are added:

$$\frac{\Gamma \Rightarrow \Delta, \varphi[t]}{\Gamma \Rightarrow \Delta, \exists x\varphi}(\Rightarrow \exists) \qquad \frac{\Gamma \Rightarrow \varphi[b]}{\Gamma \Rightarrow \Delta, \forall x\varphi}(\Rightarrow \forall)$$

$$\frac{\varphi[b], \Gamma \Rightarrow \Delta}{\exists x\varphi, \Gamma \Rightarrow \Delta}(\exists \Rightarrow) \qquad \frac{\varphi[t], \Gamma \Rightarrow \Delta}{\forall x\varphi, \Gamma \Rightarrow \Delta}(\forall \Rightarrow)$$

where t is a term and b is an *eigenvariable*, which should not occur in the conclusion. This restriction on b is called a proviso for variables. Note the form of $(\Rightarrow \forall)$ similar to $(\Rightarrow\rightarrow)$ and $(\Rightarrow \neg)$. All the rules except for $(\Rightarrow \forall)$, $(\Rightarrow\rightarrow)$ are called *invertible*. Weakening and contraction (but not cut) are rules of LJm.

15.0.1. Canonical Model, Admissibility of Cut

This section extends Section 8.2. to predicate logic. We consider possibly infinite sequents $\Gamma \Rightarrow \Delta$, where Γ, Δ are finite or countable multisets of formulas, $\Delta \neq \emptyset$. The Γ is the *antecedent*, Δ is the *succedent*. Such a sequent is *derivable* if $\Gamma' \Rightarrow \Delta'$ is derivable for some finite sets $\Gamma' \subseteq \Gamma, \Delta' \subseteq \Delta$.

A sequent S is *pure*, if no variable occurs in S both free and bound, and an infinite number of variables are absent from S (that is, these variables do not occur in S at all).

NOTE. If an infinite number of variables, say, $x_2, x_4, x_6, \ldots$ are absent from a sequent S, then S can be made pure by renaming bound occurrences of variables. Indeed rename bound variables into $x_4, x_8, x_{12}, \ldots$. Since we disregard renaming of bound variables, it is assumed that S was pure from the beginning.

DEFINITION 15.1. *If* $\mathbf{T}$ *is a set of terms, then the set* $Sub(\alpha, \mathbf{T})$ *(of* subformulas *of* α *with quantifiers instantiated by* $\mathbf{T}$*) is defined in a natural way:*

$$Sub(\alpha, \mathbf{T}) := \{\alpha\} \text{ for atomic } \alpha;$$

$$Sub(\alpha \odot \beta, \mathbf{T}) := \{\alpha \odot \beta\} \cup Sub(\alpha, \mathbf{T}) \cup Sub(\beta, \mathbf{T}) \text{ for } \odot \equiv \&, \vee, \rightarrow$$

$$Sub(\mathbf{Q}x\alpha, \mathbf{T}) := \{\mathbf{Q}x\alpha\} \cup \cup_{t \in \mathbf{T}} Sub(\alpha[x/t]) \text{ for } \mathbf{Q} \equiv \forall, \exists$$

For a set Γ *of formulas, let* $Sub(\Gamma, \mathbf{T}) := \cup_{\alpha \in \Gamma} Sub(\alpha, \mathbf{T})$.

DEFINITION 15.2. *A sequent* $\Gamma \Rightarrow \Delta$ *is* falsified *in a world* w *of a Kripke model* (W, R, D, V) *if* $V(\&\Gamma, w) = 1$, $V(\vee\Delta, w) = 0$. *This implies* $V(\Gamma \Rightarrow \Delta, w) = 0$.

In this section $\Gamma \vdash \Delta$ means that $\Gamma \Rightarrow \Delta$ is derivable in LJm.

The schema of the completeness proof is the same as for LJp (Section 8.2.). The new non-trivial step is to define an individual domain; it consists of terms.

We prove that any sequent underivable in LJm is falsified in some world of a canonical model; that is, this model is suitable for falsification of all underivable sequents.

DEFINITION 15.3. *A sequent* $\Gamma \Rightarrow \Delta$ *is* complete *with respect to a set* $\mathbf{T}$ *of terms if it is underivable and for any formula* $\phi \in Sub(\Gamma \cup \Delta, \mathbf{T})$, *either* $\phi \in \Gamma \cup \Delta$ *or:*

$$\Gamma \vdash \Delta, \phi \text{ and } \phi, \Gamma \vdash \Delta. \tag{15.1}$$

A sequent S *is* complete *if it is complete with respect to the set* $\mathcal{P}(S)$ *of all terms constructed from all free variables of* S *(and one more if* S *has no free variables) by all function symbols.*

Let $D(S) := \mathcal{P}(S)$.

DEFINITION 15.4. *A sequent* $S = \Gamma \Rightarrow \Delta$ *is* saturated for invertible rules *iff the following conditions are satisfied:*

all propositional conditions $(\Rightarrow \&) - (\rightarrow\Rightarrow)$ *from Definition 8.2.;*

$(\Rightarrow \exists)$ $\exists x\varphi \in \Delta$ *implies* $\varphi[t] \in \Delta$ *for all* $t \in D(S)$

$(\exists \Rightarrow)$ $\exists x\varphi \in \Gamma$ *implies* $\varphi[x/t] \in \Gamma$ *for some* $t \in D(S)$

$(\forall \Rightarrow)$ $\forall x\varphi \in \Gamma$ *implies* $\varphi[x/t] \in \Gamma$ *for all* $t \in D(S)$

LEMMA 15.1. *(saturation) If* $S \equiv \Gamma \Rightarrow \Delta$ *is complete with respect to* $D(S)$ *and the condition* $(\exists \Rightarrow)$ *is satisfied, then* S *is saturated for invertible rules.*

Proof. Propositional cases are treated in Lemma 8.2., and $(\exists \Rightarrow)$ is assumed. Consider remaining cases.

$(\forall \Rightarrow)$: Let $\forall x\phi \in \Gamma$. If for some $t \in D(S)$, we have $\phi[x/t] \in \Delta$ or $\Gamma \vdash \Delta, \phi[x/t]$, then $\Gamma \vdash \Delta$; hence $\phi[x/t] \in \Gamma$.

$(\Rightarrow \exists)$: Let $\exists x\phi \in \Delta$. If for some $t \in D(S)$, we have $\phi[x/t] \in \Gamma$ or $\phi, \Gamma \vdash \Delta$, then $\Gamma \vdash \Delta$; hence $\phi[x/t] \in \Delta$. $\dashv$

DEFINITION 15.5. *A set* $\mathbf{T}$ *of terms is* closed *if* $\mathbf{T}$ *is non-empty and closed under every function symbol* f*: if* $t_1, \ldots, t_n \in \mathbf{T}$ *then* $f(t_1, \ldots, t_n) \in \mathbf{T}$.

Note that $D(S)$ is closed for every sequent S.

LEMMA 15.2. *(completion) Let* $\mathbf{T}$ *be a set of terms and let* S *be an underivable sequent such that there is an infinite number of variables not occurring in* $\mathbf{T} \cup D(S)$*. Then* S *can be extended to a complete pure sequent complete with respect to* $\mathbf{T} \cup D(S)$*, closed under invertible rules and consisting of subformulas of* S.

Proof. To simplify notation assume that all predicate (function) symbols (including constants) are listed in a sequence $P_0, P_1, \ldots$ (in a sequence $f_0, f_1, \ldots$) and all variables are listed in a sequence $x_0, x_1, \ldots, y_0, y_1, \ldots$. Assume also that all free (bound) variables of $S := \Gamma_0 \Rightarrow \Delta_0$ are among $x_1, x_3, x_5, \ldots$, (among $y_0, y_1, y_2, \ldots$). Let the *index* of an expression (term, formula) be the maximum of its length and maximum subscript (of function or predicate symbol or variable) in this expression. The number of expressions with index $\leq i$ is finite.

We define inductively the sequences $\Gamma_0 \subset \Gamma_1 \subset \ldots, \Delta_0 \subset \Delta_1 \subset \ldots$ such that $\Gamma_i \Rightarrow \Delta_i$ is complete and satisfies condition $\exists \Rightarrow$ up to index i.

If Γ_i, Δ_i are already defined, let $\mathbf{T}_i$ be the set of all terms of index $\leq i+1$ constructed from free variables of $\Gamma_i \Rightarrow \Delta_i$ and let $\Phi_{i+1} \equiv \{\phi_1, \ldots, \phi_n\}$ be the set of all formulas of index $\leq i+1$ in $Sub(\Gamma_i \cup \Delta_i, \mathbf{T}_i)$. Examine in turn all formulas ϕ_j, $j = 1, \ldots, n$

Let $\Gamma_{i+1,0} := \Gamma_i$, $\Delta_{i+1,0} := \Delta_i$. If $\phi_{j+1}, \Gamma_{i+1,j} \Rightarrow \Delta_{i+1,j}$ is derivable, let $\Gamma_{i+1,j+1} := \Gamma_{i+1,j}$. If $\phi_{j+1}, \Gamma_{i+1,j} \Rightarrow \Delta_{i+1,j}$ is not derivable and ϕ_{j+1} does not begin with $\exists$, let $\Gamma_{i+1,j+1} := \phi_{j+1}, \Gamma_{i+1,j}$. If $\phi_{j+1} := \exists y\psi$, then let $\Gamma_{i+1,j+1} := \exists y\psi, \psi[y/x_{4k}], \Gamma_{i+1,j}$, where x_{4k} is the first variable of this form not free in $\Gamma_{i+1,j}, \exists y\psi$. After this let $\Delta_{i+1,j+1} := \Delta_{i+1,j}, \phi_{j+1}$ if $\Gamma_{i+1,j+1} \Rightarrow \Delta_{i+1,j+1}, \phi_{j+1}$ is underivable, and $\Delta_{i+1,j+1} := \Delta_{i+1,j}$ otherwise.

Now let $\Gamma_{i+1} := \cup_{j \leq n} \Gamma_{i+1,j}$, $\Delta_{i+1} := \cup_{j \leq n} \Delta_{i+1,j}$.

We easily prove the following claim by induction on i (with a subsidiary induction on $j = 1, \ldots, n$):

Claim. (a) The sequent $\Gamma_i \Rightarrow \Delta_i$ is not derivable.

(b) For every $\phi \in \Phi_i$ the sequent $\Gamma_i \Rightarrow \Delta_i$ is complete for ϕ, that is, either $\phi \in \Gamma_i \cup \Delta_i$ or $\Gamma_i \vdash \Delta_i, \phi$ and $\phi, \Gamma_i \vdash \Delta_i$.

(c) If $\exists y\psi \in \Phi_i$, then $\psi[y/t] \in \Gamma_i$ for some t.

Let $\Gamma := \bigcup \Gamma_i$ and $\Delta := \bigcup \Delta_i$. Now $S := \Gamma \Rightarrow \Delta$ is underivable by Claim (a). To prove that S is complete for $D(S) \equiv \cup_i \mathbf{T}_i$, take $\phi \in Sub(S, D(S))$, that is, $\phi \in Sub(\Gamma_k \cup \Delta_k, \mathbf{T}_k)$ for some k. If $i = max(index(\phi), k)$, then $\phi \in \Phi_{i+1}$, and by Claim (b), S is complete for ϕ. If in addition $\phi \equiv \exists y\psi$, then by Claim (c) $\psi[y/t] \in \Gamma$. ⊣

DEFINITION 15.6.

$$R(\Gamma \Rightarrow \Delta)(\Gamma' \Rightarrow \Delta') \quad \textit{iff} \quad \Gamma \subseteq \Gamma' \textit{ and } D(\Gamma \Rightarrow \Delta) \subseteq D(\Gamma' \Rightarrow \Delta')$$

DEFINITION 15.7. *A set M of infinite sequents is* saturated *for the domain function D if each $S \in M$ is saturated for invertible rules and M is saturated for non-invertible rules. The latter means the same conditions as in the propositional case and the following:*

($\Rightarrow \forall$) *If $(\Gamma \Rightarrow \Delta) \in M$ and $\forall x\varphi \in \Delta$, then there is a $\Gamma' \Rightarrow \Delta'$ in M with $R(\Gamma \Rightarrow \Delta)(\Gamma' \Rightarrow \Delta')$ and $\varphi[t] \in \Delta'$ for some term t.*

As in the propositional case define a Kripke model $K_M = \langle M, R, \mathcal{D}, V\rangle$, where $\mathcal{D}(S) = \mathcal{P}(S)$ and $V(f)(t_1, \ldots, t_n) := f(t_1, \ldots, t_n)$

$V(\varphi, \Gamma \Rightarrow \Delta) := 1$ iff $\varphi \in \Gamma$ for atomic φ.

DEFINITION 15.8. *Consider the* canonical *Kripke model: $\mathbf{K} = \mathbf{K}_W$ where W is the set of all complete sequents.*

We now prove that $\mathbf{K}$ falsifies every invalid formula.

LEMMA 15.3. *The set W of all complete sequents is saturated*

Proof. Only condition ($\Rightarrow \forall$) is to be checked. Assume that $\Gamma \Rightarrow \Delta, \forall x\phi$ is in W. Then the sequent $\Gamma \Rightarrow \phi[x/y]$ for a new variable y is underivable. By the completion Lemma 15.0.1, it can be extended to a complete sequent $w' = \Gamma' \Rightarrow \Delta'$ satisfying $R(\Gamma \Rightarrow \phi[x/y])(\Gamma' \Rightarrow \Delta')$ as required. ⊣

THEOREM 15.1. *Let M be a saturated set. Then for $w \equiv \Gamma \Rightarrow \Delta$, $w \in M$:*

$$\theta \in \Gamma \text{ implies } V_M(\theta, w) = 1 \tag{15.2}$$

$$\theta \in \Delta \text{ implies } V_M(\theta, w) = 0 \tag{15.3}$$

$$V_M(\Gamma \Rightarrow \Delta, w) = 0; \text{ that is, } \Gamma \Rightarrow \Delta \text{ is falsified in the world } w \text{ of } K_M\ . \tag{15.4}$$

Proof. Relation (15.4) is an immediate consequence of (15.2, 15.3), which are proved by simultaneous induction on the formula θ by the same argument as in the propositional case (Theorem 8.2.).

Consider only induction step for quantifier rules.

$\exists$: $\theta = \exists x\phi$. If $\theta \in \Gamma$, then $\phi[t] \in \Gamma$ for some t by the saturation for invertible rules, so $V_M(\phi[t], w) = 1$ by the induction hypothesis; hence $V_M(\theta, w) = 1$ by the truth condition for $\exists$.

If $\theta \in \Delta$, then $\phi[t] \in \Delta$ for all $t \in D(\Gamma \Rightarrow \Delta)$ by the saturation condition. Hence $V_M(\phi[t], w) = 0$ by the induction hypothesis, so $V_M(\theta, w) = 0$.

$\forall$: $\theta = \forall x\phi$. If $\theta \in \Gamma$, then for every $w' \equiv \Gamma' \Rightarrow \Delta'$ such that $R_M\, w\, w'$ (which implies $\Gamma \subseteq \Gamma'$), we have $\theta \in \Gamma'$. By the saturation condition, this implies $\phi[t] \in \Gamma'$ for all $t \in D(w')$. By the induction hypothesis, $V_M(\phi[t], w') = 1$. This implies $V_M(\forall x\phi, w) = 1$.

If $\theta \in \Delta$, then by the saturation condition ($\Rightarrow \forall$), we have $w' \equiv \Gamma' \Rightarrow \Delta'$ in M such that $R_M\, w\, w'$, $\phi[t] \in \Delta'$. By the induction hypothesis, $V(\phi[t], w') = 0$, which implies $V(\forall x\phi, w) = 0$, as required. ⊣

COROLLARY 15.1. *(completeness) Each sequent underivable in LJm is falsified in the canonical model* **K**. *Hence every valid sequent is derivable in LJm.*

Proof. By the completion Lemma 15.0.1. any underivable sequent $\Gamma \Rightarrow \Delta$ can be extended to a complete sequent $w = \Gamma' \Rightarrow \Delta'$; by the previous Theorem $V(\Gamma' \Rightarrow \Delta', w) = 0$. Hence $V(\Gamma \Rightarrow \Delta, w) = 0$ by monotonicity. ⊣

THEOREM 15.2. *(soundness and completeness). A formula is derivable in* LJm *iff it is valid, iff it is valid in all pointed models where accessibility relation is a partial order.*

Proof. First apply the previous Lemma and Corollary 15.0.1(a), then apply Lemma 14.1.. ⊣

COROLLARY 15.2. *Cut rule is admissible in LJm, and LJm is equivalent to NJ with respect to derivability of sequents* $\Gamma \Rightarrow \alpha$.

Proof. Exactly as for Theorem 8.2.. ⊣

15.1. Translation into the Classical Logic

Translation of a formula ϕ into a predicate formula expressing the validity of ϕ given in the Section 8.3. is extended to predicate logic in a natural way. To every n-ary function or predicate symbol J we assign an $(n+1)$-ary symbol J' of the same type. We also choose new binary predicate symbols R, D and a new unary predicate symbol W.

DEFINITION 15.9. *For any term t, any predicate formula φ, and individuum variable w, we define a term (t, w) and a predicate formula (φ, w) by induction on t, φ.*

$$\begin{array}{lcl}
(x, w) & := & x \\
(f(t_1, \ldots, t_n), w) & := & f'((t_1, w), \ldots, (t_n, w), w) \\
(P(t_1, \ldots, t_n), w) & := & P'((t_1, w), \ldots, (t_n, w), w) \\
(\varphi \& \psi, w) & := & (\varphi, w) \& (\psi, w) \\
(\varphi \vee \psi, w) & := & (\varphi, w) \vee (\psi, w) \\
(\varphi \rightarrow \psi, w) & := & \forall w'(W(w') \& R(w, w') \& (\varphi, w') \rightarrow (\psi, w')) \\
(\exists x \phi, w) & := & \exists x(D(x, w) \& (\phi, w)) \\
(\forall x \phi, w) & := & \forall w' \forall x(W(w') \& R(w, w') \& D(x, w') \rightarrow (\phi, w'))
\end{array} \tag{15.5}$$

where w' is a new variable.

List all free variables, function symbols, and predicate symbols of the formula ϕ. To simplify notation, assume that there are just two free variables v_1, v_2, two n-ary predicate symbols f_1, f_2, and two n-ary predicate symbols P_1, P_2. Choose distinct new variables $w, w_1, w_2, x, x_1, \ldots, x_n, a$, and denote $\mathbf{x} := x_1, \ldots, x_n$. Define

$$\varphi^P := \kappa \rightarrow (\phi, a)$$

where:

$$\begin{gathered}\kappa := W(a)\& \\ \&_{i\leq 2}D(v_i,a)\&\&_{i\leq 2}\forall x_1\ldots\forall x_n(\&_{j\leq n}D(x_j,w)\rightarrow D(f_i'(\mathbf{x},w),w))\& \\ \forall w(W(w)\rightarrow R(w,w))\& \\ \forall w\forall w_1\forall w_2(W(w)\&W(w_1)\&W(w_2)\&R(w,w_1)\&R(w_1,w_2)\rightarrow R(w,w_2))\& \\ \forall w\forall w'(W(w)\&W(w')\&R(w,w')\rightarrow[\forall x(D(x,w)\rightarrow D(x,w'))\& \\ \&_{i\leq 2}\forall x_1\ldots\forall x_n(\&_{j\leq n}D(x_j,w)\rightarrow f_i'(\mathbf{x},w)=f_i'(\mathbf{x},w'))\& \\ \&_{i\leq 2}\forall x_1\ldots\forall x_n(\&_{j\leq n}D(x_j,w)\rightarrow P_i'(\mathbf{x},w)\rightarrow P_i'(\mathbf{x},w'))])\end{gathered}$$

THEOREM 15.3. *A formula φ is derivable in LJm iff φ^P is derivable in the classical predicate calculus with equality.*

Proof. Extend the proof of Theorem 8.3.. A first-order model:

$$M = \langle f_1', f_2', W, D, R, P_1', P_2', a_0, a_1, a_2\rangle$$

with an interpretation for the functions, predicates, and the constant in φ^P, satisfying the premise κ of φ^P generates a Kripke model:

$$M' = \langle W, R, D, V_{M'}\rangle,$$

where $V_{M'}(P_i)(\mathbf{d},w)=1$ iff $M\models P_i'(\mathbf{d},w)$ for every $w\in W$ and $\mathbf{d}:=d_1,\ldots,d_n$ with $M\models D(d_1,w)\&\ldots\&D(d_n,w)$.

In the other direction, any Kripke model M' with a distinguished world a_0 generates a first-order model M by means of (15.5), that is, $M\models P'(\mathbf{d},w)$ iff $M',w\models P(\mathbf{d})$ and $f'(\mathbf{d},w)=V(f)(\mathbf{d},w)$. Moreover by induction on ψ with variables $v_1,\ldots,v_m$, it is easy to prove that:

$$M\models\psi^P[a/w,v_1/d_1,\ldots,v_m/d_m]\quad\text{iff}$$

$$V_{M'}(\psi[v_1/d_1,\ldots,v_m/d_m],w)=1 \text{ and } w\in W,\ d_1,\ldots,d_n\in D(w).$$

So φ^P is valid (in the first-order logic) iff φ is valid intuitionistically. ⊣

15.2. System LJ

A one-succedent version LJ of LJm is obtained by restricting the succedent to one formula in LJm. In other words, sequents take the form $\Gamma\Rightarrow\alpha$, axioms and propositional rules are the same as in LJp, and quantifier rules are as follows:

Systen LJ: Quantifier rules

$$\frac{\Gamma\Rightarrow\phi[x/t]}{\Gamma\Rightarrow\exists x\phi}\qquad\frac{\phi[x/y],\Gamma\Rightarrow\theta}{\exists x\phi,\Gamma\Rightarrow\theta}$$

$$\frac{\phi[x/t],\Gamma\Rightarrow\theta}{\forall x\phi,\Gamma\Rightarrow\theta}\qquad\frac{\Gamma\Rightarrow\phi[x/y]}{\Gamma\Rightarrow\forall x\phi}$$

15.2.1. Translating LJpm into LJp

LEMMA 15.4. *(soundness) All rules of LJm are derivable in NJ under disjunctive translation of* $\Gamma \Rightarrow \Delta$ *into* $\Gamma \Rightarrow \vee\Delta$. *Hence LJm is sound: Every derivable sequent is valid.*

Proof. Axioms and propositional rules are covered by Lemma 8. Consider quantifier rules. Succedent rules of LJm go into introduction rules of NJ plus $\vee I$. Antecedent rules follow from NJ-derivability of $\forall x\phi \Rightarrow \phi[x/t]$, $\forall y(\phi[x/y]\&\gamma \rightarrow \delta) \rightarrow (\exists x\phi\&\gamma \rightarrow \delta)$. ⊣

A disjunctive translation works as in Section 11.2.

LEMMA 15.5. *(a)* $\Gamma \Rightarrow \Delta$ *is derivable in LJm iff* $\Gamma \Rightarrow \vee\Delta$ *is derivable in LJ + cut.*

(b) In particular, LJpm is equivalent to LJp + cut for sequents $\Gamma \Rightarrow \alpha$.

⊣

Pruned derivation in LJm is defined exactly as before (Definition 11.3.), and the following two Lemmas are established exactly as Lemmas 11.3. and 11.3.

LEMMA 15.6. *Every derivation* $d : \Gamma \Rightarrow \Delta$ *in LJm can be transformed into a pruned derivation* $d' : \Gamma \Rightarrow \Delta$ *by moving weakenings downward and deleting occurrences of formulas and whole branches.*

⊣

LEMMA 15.7. *If* $d : \Gamma \Rightarrow \Delta$ *is a pruned deduction in LJpm from arbitrary 1-sequents containing no* $\vee \Rightarrow$*-inferences, then* $\Gamma \Rightarrow \Delta$ *is a 1-sequent up to weakenings.*

⊣

In the definition of a movable rule, a proviso for $(\forall \Rightarrow)$ is added.

DEFINITION 15.10. *Let* $d : \Gamma \rightarrow \Delta$ *be a derivation in LJpm and let L be an antecedent rule in d with a principal formula* π *and conclusion* $\pi, \Gamma' \Rightarrow \Delta'$. *Then L is* movable *if* $\pi, \Gamma' \subseteq \Gamma$ *and:*

(a) If $L \equiv \rightarrow\Rightarrow$, *then* Δ' *is pruned from the left premise;*

(b) if $L \equiv \forall \Rightarrow$:

$$\frac{\phi[x/t], \Gamma' \Rightarrow \Delta'}{\forall x\phi, \Gamma' \Rightarrow \Delta'}\ L$$
$$\vdots$$
$$\forall x\phi, \Gamma \Rightarrow \Delta$$

then no variable occurring in t is an eigenvariable of an $\exists$*-inference below L.*

After this the proof of Lemma 11.3. does not change.

LEMMA 15.8. *Let $d : \Gamma \Rightarrow \Delta$ be a pruned deduction in LJm from 1-sequents containing an $\vee \Rightarrow$-rule but no non-invertible rules. Then d contains a* movable *rule.*

⊣

The proof of Theorem 11.3. also extends easily. Some eigenvariables should be renamed to make possible permutation of movable rules.

THEOREM 15.4. *Every derivation of 1-sequent $\Gamma \Rightarrow \alpha$ in LJm can be transformed by renaming eigenvariables, permuting movable rules, and pruning into a derivation in LJ.*

Proof. We use induction on the number of logical inferences in the given derivation. If all premises of the last rule are 1-sequents (up to weakenings), then apply IH. Otherwise delete from the given derivation all sequents above the lowermost 1-sequent (up to pruning) in every branch. We are left with a deduction *d* from 1-sequents by invertible rules, since conclusions of non-invertible rules are 1-sequents. If this deduction does not contain $\vee \Rightarrow$-inferences, use Lemma 15.2.1. Otherwise there is a movable rule. Moving it down and applying IH to its premises concludes the proof. Let us consider quantifier case.

Case 1. There is a movable $\forall \Rightarrow$-inference. Then *d* is of the form shown on the left below. Permute it down as shown on the right, and IH can be applied to *d'*. The permutation is made possible by the proviso in Definition 15.2.1(b).

$$\dfrac{\dfrac{\vdots}{\alpha[t], \Sigma' \Rightarrow \Delta}}{\forall x\alpha, \Sigma' \Rightarrow \Delta}\,L \quad \begin{matrix} | \\ \forall x\alpha, \Sigma \Rightarrow \gamma \end{matrix} \qquad\qquad \dfrac{\dfrac{\begin{matrix} | \\ \alpha[t], \Sigma' \Rightarrow \Delta \\ | \\ d' : \alpha[t], (\forall x\alpha)^0, \Sigma \Rightarrow \gamma \end{matrix}}{\forall x\alpha, (\forall x\alpha)^0, \Sigma \Rightarrow \gamma}}{\forall x\alpha, \Sigma \Rightarrow \gamma}\,contr$$

where the contaction rule and $(\forall x\alpha)^0$ are present only if the lowermost $\forall x\alpha$ has predecessors other than the principal formula of L.

Case 2. There is a movable $\exists \Rightarrow$-inference. This case is similar to the previous case:

$$\dfrac{\dfrac{\vdots}{\alpha[b], \Sigma' \Rightarrow \Delta}}{\exists x\alpha, \Sigma' \Rightarrow \Delta} \quad \begin{matrix} | \\ \exists x\alpha, \Sigma \Rightarrow \gamma \end{matrix} \qquad\qquad \dfrac{\dfrac{\begin{matrix} | \\ \alpha[b], \Sigma' \Rightarrow \Delta \\ | \\ \alpha[b], (\exists x\alpha)^0, \Sigma \Rightarrow \gamma \end{matrix}}{\exists x\alpha, (\exists x\alpha)^0, \Sigma \Rightarrow \gamma}}{\exists x\alpha, \Sigma \Rightarrow \gamma}$$

This concludes the proof. ⊣

15.3. Interpolation Theorem

In this section assume that the language of predicate logic does not contain function symbols, so that the terms are just variables.

If E is a formula or sequent, then L_E stands for the list of predicate symbols occurring in E plus $\{\top, \bot\}$.

A *Craig interpolant* ι for an implication $\beta \to \alpha$ and for a *partition* $\Gamma; \Delta \Rightarrow \alpha$ of a sequent $\Gamma, \Delta \Rightarrow \alpha$ is defined exactly as in the propositional case (Section 12) by the properties:

$$\vdash \beta \to \iota \quad \vdash \iota \to \alpha \qquad \text{and} \quad L_\iota \subseteq L_\alpha \cap L_\beta$$
$$\vdash \Gamma \Rightarrow \iota, \quad \vdash \iota, \Delta \Rightarrow \alpha \qquad \text{and} \quad L_\iota \subseteq L_\Gamma \cap L_{\Delta \Rightarrow \alpha}$$

NOTE. By prefixing quantifiers to a Craig interpolant ι for $\beta \to \alpha$, it is possible to achieve $FV(\iota) \subseteq FV(\alpha) \cap FV(\beta)$. Indeed if x is free in α, but not in β, and y is free in β, but not in α, then $\vdash \beta \to \forall x \exists y \iota$, $\vdash \forall x \exists y \iota \to \alpha$.

THEOREM 15.5. *(interpolation theorem) (a) If $\Gamma, \Delta \Rightarrow \alpha$ is derivable, then there is a Craig interpolant for $\Gamma; \Delta \Rightarrow \alpha$.*

(b) If $\alpha \to \beta$ is derivable, then there is a Craig interpolant for $\alpha \to \beta$.

Proof. Part (b) immediately follows from Part (a). Proof of Part (a) by induction on a given derivation $d : \Gamma, \Delta \Rightarrow \alpha$ in LJ is an extension of the proof of Theorem 12. Axioms and propositional rules are treated exactly as in the propositional case. Consider quantifier rules of LJ.

Case 1. The d ends in a quantifier rule L without eigenvariable (that is, $L \equiv \Rightarrow \exists$ or $L \equiv \forall \Rightarrow$). Then the interpolant ι for the premise is preserved for the conclusion. If for example $L \equiv \Rightarrow \exists$, then:

$$\frac{\Gamma, \Delta \Rightarrow \alpha[x/t]}{\Gamma, \Delta \Rightarrow \exists x \alpha}\,L \qquad \begin{array}{lll} & \iota, \Delta \Rightarrow \alpha[x/t] & \textit{by IH} \\ \Gamma \Rightarrow \iota & \iota, \Delta \Rightarrow \exists x \alpha & \textit{by} \Rightarrow \exists \end{array}$$

Case 2. The d ends in a quantifier rule L with an eigenvariable y (that is, $L \equiv \Rightarrow \forall$, $\exists \Rightarrow$). Then an interpolant ι for the premise is prefixed by a quantifier $\exists y$ if the principal formula of L is in Γ. The interpolant is prefixed by $\forall y$ if the principal formula of L is in $\Delta \Rightarrow \alpha$; for example:

$$\frac{\Gamma, \Delta \Rightarrow \phi[x/y]}{\Gamma, \Delta \Rightarrow \forall x \phi} \qquad \begin{array}{ll} & \iota, \Delta \Rightarrow \phi[x/y] \\ \Gamma \Rightarrow \iota & \forall y \iota, \Delta \Rightarrow \phi[x/y] \\ \Gamma \Rightarrow \forall y \iota & \forall y \iota, \Delta \Rightarrow \forall x \phi \end{array}$$

Proviso for eigenvariable in the rule introducing $\forall y \iota$ into succedent is satisfied, since $y \notin FV(\Gamma)$. ⊣

Now the Beth definability Theorem for predicate logic is obtained similarly to the propositional case. Let:

$$\mathbf{x} := x_1, \ldots, x_n \qquad \forall \mathbf{x} := \forall x_1 \ldots \forall x_n$$

DEFINITION 15.11. *A formula α* implicitly defines *an n-ary predicate symbol P iff the formula:*

$$\alpha \& \alpha[P/Q] \to \forall \mathbf{x}(P\mathbf{x} \leftrightarrow Q\mathbf{x}) \tag{15.6}$$

is derivable for a new n-ary predicate symbol Q and new variables $\mathbf{x}$.

Formula (15.6) says that at most one predicate P (up to equivalence) satisfies α.

THEOREM 15.6. *(Beth definability theorem). If a formula α implicitly defines P, then α explicitly defines P: There exists a formula ι with $FV(\iota) \subseteq \mathbf{x} \cup FV(\alpha)$ that does not contain P such that:*

$$\vdash \alpha \to \forall \mathbf{x}(P\mathbf{x} \leftrightarrow \iota)$$

Proof. Derivability of (15.6) implies derivability of the sequent:

$$P\mathbf{x}, \alpha, \alpha[P/Q] \Rightarrow Q\mathbf{x}$$

Take ι to be an interpolant for the partition $P\mathbf{x}, \alpha;\ \alpha[P/Q] \Rightarrow Q\mathbf{x}$ with $FV(\iota) \subseteq \mathbf{x} \cup FV(\alpha)$:

$$P\mathbf{x}, \alpha \Rightarrow \iota; \qquad \iota, \alpha[P/Q] \Rightarrow Q\mathbf{x}.$$

This implies (after substituting Q by P in the second sequent):

$$\alpha \Rightarrow P\mathbf{x} \to \iota \qquad \alpha \Rightarrow \iota \to P\mathbf{x}$$

as required. ⊣

Chapter 16

Proof-Search in Predicate Logic

It is possible to extend the treatment of proof-search for LJpm* to the predicate case without essential change, but certain technical complications arise. We introduce instead an important technique due to Kripke.

Derivable objects of the system LJm* are tableaux $T \equiv S_1 * \ldots * S_n$ constructed from the sequents S_i in the language of predicate logic and enriched by a binary relation r on the set $\{1, \ldots, n\}$ called *immediate accessibility relation.* The rij always implies $i < j$; R denotes the *reflexive transitive closure* of r:

$$Rij \text{ iff } i = j \text{ or there are } i_0 = i, i_1, \ldots i_n = j \text{ such that } ri_k i_{k+1} \text{ for all } k < n$$

Axioms and propositional inference rules of the system LJm* are defined exactly as for the system LJpm* (Section 11.1.). Relation r is the same for premises and conclusion of every *invertible rule*: $\Rightarrow \&, \& \Rightarrow, \Rightarrow \vee, \vee \Rightarrow, \rightarrow\Rightarrow$ as well as quantifier rules $\exists \Rightarrow, \forall \Rightarrow, \Rightarrow \exists$, and a *transfer* rule that follows. For a *quasi-invertible* rule ($\Rightarrow\rightarrow$ or $\Rightarrow \forall$ below) analyzing a component S_i in the conclusion and generating a new component S_{n+1} in the premise, the relation r for the premise is obtained by adding $ri(n+1)$ to the relation r for the conclusion. Quantifier rules are defined in a natural way, and there is an additional invertible *transfer* rule:

$$\frac{T * \phi, \Gamma \Rightarrow \Delta * T' * \phi, \Sigma \Rightarrow \Pi * T''}{T * \phi, \Gamma \Rightarrow \Delta * T' * \Sigma \Rightarrow \Pi * T''}\, transfer \qquad \begin{matrix} \text{if} & S_i \equiv \phi, \Gamma \Rightarrow \Delta, \\ rij & S_j \equiv \Sigma \Rightarrow \Pi \end{matrix}$$

$$\frac{T * \Gamma \Rightarrow \Delta, \varphi[x/t], \exists x\varphi * T'}{T * \Gamma \Rightarrow \Delta, \exists x\varphi * T'}(\Rightarrow \exists)$$

$$\frac{T * \Gamma \Rightarrow \Delta, \forall x\varphi * T' * \Gamma \Rightarrow \varphi[x/y]}{T * \Gamma \Rightarrow \Delta, \forall x\varphi * T'}(\Rightarrow \forall)$$

$$\frac{T * \varphi[x/y], \Gamma \Rightarrow \Delta * T'}{T * \exists x\varphi, \Gamma \Rightarrow \Delta * T'}(\exists \Rightarrow) \qquad \frac{T * \varphi[x/t], \forall x\varphi, \Gamma \Rightarrow \Delta * T'}{T * \forall x\varphi, \Gamma \Rightarrow \Delta * T'}(\forall \Rightarrow)$$

where t is a term, y is an *eigenvariable* that should not be free in any component of the conclusion (proviso for variables).

The $FV(T)$ denotes the set of free variables of components of T.

Note that all rules of LJm* (even $\Rightarrow\rightarrow, \Rightarrow\forall$) are invertible. Our terminology reminds of non-invertible ancestors of these rules in LJm. With respect to equivalence of LJm* and LJm, note that neither pruning nor translating $S_1 * \ldots * S_n$ as $S_1 \vee \ldots \vee S_n$ work for the transfer rule.

For a tableau $T \equiv S_1 * \ldots * S_n$ with a binary relation r, we define, following S. Kripke, a *characteristic formula* $\chi(S_i;T)$ of the component $S_i \equiv \Gamma \Rightarrow \Delta$ by induction on $n-i$. Let $v_1, \ldots, v_m$ $(m \geq 0)$ be the list of all free variables of S_i that are not free in any S_k with $Rki,\ k \neq i$. Let $j_1, \ldots, j_l$ $(l \geq 0)$ be the list of all j with rij. Then let

$$\chi(S_i;T) := \forall v_1 \ldots \forall v_m (\&\Gamma \rightarrow \vee\Delta \vee \chi(S_{j_1};T) \vee \ldots \vee \chi(S_{j_l};T))$$

$$\chi(T) := \chi(S_1;T)$$

LEMMA 16.1. *(a) Every rule of LJm* is derivable under χ-translation in LJm with cut:*

$$\chi(T') \vdash \chi(T) \text{ in LJm for every one-premise rule } T'/T \tag{16.1}$$

$$\chi(T_0), \chi(T_1) \vdash \chi(T) \text{ in LJm for every two-premise rule } T_0, T_1/T. \tag{16.2}$$

(b) Every sequent derivable in LJm is derivable in LJm+cut.*

Proof. Part (b) follows from Part (a) since $\chi(S;S) \vdash S$ in LJm+cut.
To prove Part (a), we establish that:

$$\chi(S_i^0;T_0), \chi(S_i^1;T_1) \vdash \chi(S_i;T) \tag{16.3}$$

in LJm for every two-premise rule $T_0, T_1/T$ with the principal component S_i in the conclusion T and side components S_i^0, S_i^1 in the premises T_0, T_1, and similar relations (see below) for one-premise rules. Note that (up to redundant $\forall$ -quantifiers):

$$\chi(T) \equiv \theta[p/\chi(S_i;T)] \qquad \chi(T_l) \equiv \theta[p/\chi(S_i^l;T)],\ l = 0, 1$$

where p has exactly one occurrence in θ, and this occurrence is not in a scope of any & or in a premise of any $\rightarrow$. For such p and any formulas α, β we easily prove by induction on θ:

$$\alpha, \beta \vdash \gamma \text{ implies } \theta[p/\alpha], \theta[p/\beta] \vdash \theta[p/\gamma] \tag{16.4}$$

With (16.3) this proves (16.2).

For the one-premise rule T'/T with (the leftmost) active component S_i in the conclusion T, we prove a similar relation:

$$\chi(S_i';T') \vdash \chi(S_i;T) \tag{16.5}$$

and argue as before with a slight change for $\Rightarrow \forall$ and *transfer*. We check only three rules, leaving the rest to the reader.

$\Rightarrow \&$ with the active component $S_i \equiv \Gamma \Rightarrow \Delta, \alpha_0 \& \alpha_1$. We have:

$$\chi(S_i; T) \equiv \&\Gamma \Rightarrow \vee\Delta \vee (\alpha_0 \& \alpha_1) \vee \theta$$

$$\chi(S_i^l; T') \equiv \&\Gamma \Rightarrow \vee\Delta \vee \alpha_l \vee \theta$$

and (16.3) is obvious.

$\Rightarrow \forall$ with the active component $S_i \equiv \Gamma \Rightarrow \Delta, \forall x\phi$ adding to the premise a component $S_{n+1} \equiv \Gamma \Rightarrow \phi[x/y]$. We have:

$$\chi(S_i; T) \equiv \&\Gamma \Rightarrow \vee\Delta \vee \forall x\phi \vee \theta$$

$$\chi(S_i'; T') \equiv \&\Gamma \Rightarrow \vee\Delta \vee \forall x\phi \vee \theta \vee \forall y\phi[x/y]$$

and (16.5) is obvious.

For *transfer* T'/T with active components S_i, S_j in the conclusion and $S_i \equiv \phi, \Gamma \Rightarrow \Delta$, $S_j' \equiv \phi, \Sigma \Rightarrow \Pi$ in the premise, we have (up to associativity and commutativity of disjunction):

$$\chi(S_i, T) \equiv \phi \& \&\Gamma \Rightarrow \vee\Delta \vee (\&\Sigma \Rightarrow \vee\Pi \vee \zeta) \vee \theta$$

$$\chi(S_i, T') \equiv \phi \& \&\Gamma \Rightarrow \vee\Delta \vee (\phi \& \&\Sigma \Rightarrow \vee\Pi \vee \zeta) \vee \theta$$

and (16.5) is easy to prove. ⊣

THEOREM 16.1. *The LJm* is equivalent to LJm*

Proof. Exactly as Theorem 10.1. ⊣

For every sequent S, let S_a, S_s be its antecedent and succedent:

$$\text{if } S \equiv \Gamma \Rightarrow \Delta \text{ then } S_a := \Gamma,\ S_s := \Delta$$

A *proof-search procedure* for LJm* consists of *tree extension steps*, as described in Section 10.2.: One of the leaf nodes (*goals*) T of the *proof-search tree* is analyzed by a bottom-up application of one of the rules of LJm* (including quantifier rules and *transfer*). A tableau is *closed* if it is an axiom of LJm*; a proof-search tree is *closed* if all goals (leaf nodes) are closed. A closed tree is a derivation in LJm*. The notion of a terminal node is not as important as in the propositional case: If the potential supply of terms to be substituted is infinite and one of the rules $\forall \Rightarrow, \Rightarrow \exists$ is to be applied, this supply is never exhausted. On the other hand, restrictions ensuring finiteness of this supply determine an important decidable subclass.

Let us define a complete *proof-search strategy* for LJm*. Principal formulas of all invertible rules are preserved, in particular $\exists \Rightarrow$ has the form:

$$\cfrac{\cfrac{T_1 * \phi[x/y], \exists x\phi, \Gamma \Rightarrow \Delta * T_2}{T_1 * \exists x\phi, \exists x\phi, \Gamma \Rightarrow \Delta * T_2}}{T_1 * \exists x\phi, \Gamma \Rightarrow \Delta * T_2}$$

The tree extension steps are done in groups called *stages*. A stage analyzes a goal T into a finite sequence of new goals:

$$\frac{T_1 \ldots T_n}{T}$$

A formula α that is an antecedent or succedent term in one of the components S_i of a tableau T is *passive* in T with respect to a set M of terms if all propositional conditions ($\Rightarrow$ &)-$\rightarrow\Rightarrow$ of Definition 8.2. as well as the condition ($\exists \Rightarrow$) of Definition 15.0.1for α and S_i; condition ($\forall \Rightarrow$) of Definition 15.0.1is satisfied for α, S_i and all terms $t \in M$; and conditions ($\Rightarrow\rightarrow, \Rightarrow \forall$) of Definition 15.0.1. are satisfied for α, S_i and some component S_j with rij:

$$(\Rightarrow\rightarrow): \text{ If } \alpha \equiv \phi \rightarrow \psi \in S_{is}, \text{ then } \phi \in S_{ja}, \psi \in S_{js} \text{ for some } j \text{ satisfying } rij.$$

$$(\Rightarrow \forall): \text{ If } \alpha \equiv \forall x\varphi \in S_{is}, \text{ then } \phi[x/t] \in S_{js} \text{ for some } j \text{ satisfying } rij$$

and some term t.

The active length is defined now with respect to a set of terms M:

$$alength(S_i; T, M) :=$$

$$\sum\{length(\alpha) : \alpha \text{ is a non-passive formula in } S_i \text{ w.r.t. } M\}$$

Let a tableau $T \equiv S_1 * \ldots * S_n$ with immediate accessibility relation r and a non-passive formula α in a component S_i be fixed. The *tree extension step* analyzing α with respect to a finite set M of terms is defined as in Section 10.2. if α is not an antecedent $\forall$-formula nor succedent $\exists$-formula. For a bottom-up application of rules $\exists \Rightarrow, \Rightarrow \forall$ having eigenvariables we choose a new variable. Invertible rules do not change the relation r. With a bottom-up application of a non-invertible rule ($\Rightarrow\rightarrow, \Rightarrow \forall$) to a component S_i, adding a component S_{n+1}, we extend the relation r adding $ri(n+1)$. Finally a bottom-up application of $\forall \Rightarrow, \Rightarrow \exists$ adds in one step all instances in the set M that are not already present:

$$\frac{T_1 * \phi[x/t_1], \ldots, \phi[x/t_n], \forall x\phi, \Gamma \Rightarrow \Delta * T_2}{T_1 * \forall x\phi, \Gamma \Rightarrow \Delta * T_2} \quad \begin{array}{c}\text{for all } t_k \in M \\ \text{such that } \phi[x/t_k] \notin \Gamma\end{array}$$

$$\frac{T_1 * \Gamma \Rightarrow \Delta, \phi[x/t_1], \ldots, \phi[x/t_n], \exists x\phi * T_2}{T_1 * \Gamma \Rightarrow \Delta, \exists x\phi * T_2} \quad \begin{array}{c}\text{for all } t_k \in M \\ \text{such that } \phi[x/t_k] \notin \Delta\end{array}$$

A tree extension *stage* number l for a tableau $T \equiv S_1 * \ldots * S_n$ with an immediate accessibility relation r applies a tree extension step to every antecedent or succedent formula α that is non-passive in every component S_i with respect to the set M_i of all terms with an index $\leq l$ constructed from:

$$\cup_{j:Rij} FV(S_j) \tag{16.6}$$

These extension steps can be performed in any order, but only formulas already present and active in T should be analyzed: Formulas added at a given stage are analyzed at the next stage.

We assume that there is at least one constant; otherwise, we should ensure that $D(S_1) \neq \emptyset$ by placing a variable there.

At stage 0 the tree Tr_{S_1} for a sequent S_1 consists only of S_1 with the empty (identically false) relation r. At the end of every stage, all possible non-redundant *transfer* inferences are applied bottom-up.

DEFINITION 16.1. *[terminal leaves] A leaf of the proof-search tree is* terminal, *if the tableau T situated at this node generates saturated set of sequents $S_1, \ldots, S_n$.*

Let T be a proof-search tree, and $\mathcal{B}$ be a branch of the tree, that is, a sequence:

$$T^0, T^1, \ldots \tag{16.7}$$

of tableaux such that T^0 is the goal sequent (the root of the tree) and for every i, the tableau T^{i+1} (if it exists) is a premise of a rule with conclusion T^i. Say that $\mathcal{B}$ is *non-closed* if it is infinite or its last tableau is saturated as a set of sequent. In the latter case, we treat the branch as infinite by repeating the last tableau.

For a branch $\mathcal{B}$ of a proof-search tree, every k not exceeding the length of the branch and every i denote by S^i_k the kth component of the tableau T_i. In the case when $\mathcal{B}$ is non-closed, define the kth (infinite) sequent S^∞_k of $\mathcal{B}$ by accumulating antecedents and succedents of sequents S^i_k:

$$S^\infty_k = \bigcup_i S^i_{ka} \Rightarrow \bigcup_i S^i_{ks}$$

Define binary relation r^∞ on natural numbers: $r^\infty kl$ iff rkl holds for tableau T^i for some i. R^∞ denotes reflexive transitive closure of r^∞.

Define:

$$D(S^\infty_k) := \text{ the set of all terms, constructed from } \cup_{l:Rlk} FV(S^\infty_l)$$

by function symbols.

The next Lemma shows that every meaningful application of any possible rule is eventually made in every non-closed branch of a proof-search tree.

LEMMA 16.2. *Under our strategy, the following fairness conditions are satisfied for every k, i:*

$(\& \Rightarrow)$: *If $\varphi \& \psi \in S^i_{ka}$, then $\varphi, \psi \in S^j_{ka}$ for some j.*

$(\Rightarrow \&)$: *If $\varphi \& \psi \in S^i_{ks}$, then $\varphi \in S^j_{ks}$ or $\psi \in S^j_{ks}$ for some j.*

$(\Rightarrow \vee, \vee \Rightarrow, \rightarrow \Rightarrow)$: *Similarly.*

$(\forall \Rightarrow)$: *If* $(\forall x\varphi) \in S^i_{ka}$, *then:*

(a) $\varphi[t] \in S^j_{ka}$ *for some* j *and some* t

(b) for any l *with* $R\,l\,k$ *and for every term* $t \in M_i$ *occurring in* $S^{i'}_l$ *with* $i' \leq i$, *there is a* j *such that* $\varphi[x/t] \in S^j_{ka}$.

$(\Rightarrow \exists)$: *Similarly.*

$(\Rightarrow\rightarrow)$: *If* $(\varphi \rightarrow \psi) \in S^i_{ks}$, *then there are* j, l *with* $R\,k\,l$ *such that* $\varphi \in S^j_{la}$ *and* $\psi \in S^j_{ls}$.

$(\exists \Rightarrow)$: *If* $(\exists x\varphi) \in S^i_{ka}$, *then there is a* t *such that* $\varphi[x/t] \in S^j_{ka}$ *for some* j.

$(\Rightarrow \forall)$: *If* $(\forall x\varphi) \in S^i_{ks}$, *then there are* j, l *with* $R\,k\,l$ *such that* $\varphi[x/t] \in S^j_{ls}$ *for some* t.

(Transfer): *If* $\varphi \in S^i_{ka}$, *then for every* l *with* $r^\infty(k,l)$ *there is a* j *such that* $\varphi \in S^j_{la}$.

THEOREM 16.2. *If* $\mathcal{B}$ *is a non-closed branch of a proof-search tree* Tr_{S_1} *under our strategy, then the set of infinite sequents* $S_1^\infty, S_2^\infty, \ldots$ *determined by* $\mathcal{B}$ *with the domain function* D *is saturated.*

Proof. Note that:

$$R^\infty km \text{ implies } S^\infty_{ka} \subset S^\infty_{ma}; \qquad D(S^\infty_k) \subset D(S^\infty_m) \tag{16.8}$$

The first relation follows from Lemma 16 **Transfer**. The second relation follows from the transitivity of R^∞: If $R^\infty km$, then the terms in $D(S^\infty_m)$ are constructed from $\cup_{l:R^\infty lm} FV(S^\infty_l) \supseteq \cup_{l:R^\infty lk} FV(S^\infty_l)$.

Hence $R^\infty kl$ implies accessibility between infinite sequents S^∞_k, S^∞_l. It is easy to prove closure under the invertible rules for the propositional connectives.

If for example $\varphi\&\psi \in S_{ka}$, then $\varphi\&\psi \in S^i_{ka}$ for some i. Threfore $\varphi, \psi \in S^j_{ka}$ for some j, and $\varphi, \psi \in S_{ka}$, as required.

Let $\exists x\varphi \in S_{ks}$ and $t \in D(S_k)$. Then $\exists x\varphi \in S^i_{ks}$ for all $i \geq i_0$ (since the main formula $\exists x\varphi$ is preserved in the rule $(\Rightarrow \exists)$) and $t = 0$ or t occurs in S_l for some l satisfying Rlk. The latter means that t occurs in $D(S^j_l)$ for some j. Taking the maximum of i, j we have (by fairness for the rule $\Rightarrow \exists$) that $\varphi[t] \in S^{j'}_k$ for some j', and hence $\varphi \in S_k$, as required.

The case $\forall x\varphi \in S_{ka}$ is treated similarly.

Consider the non-invertible rules. Let $(\neg\varphi) \in S_{ks}$. Then $(\neg\varphi) \in S^i_{ks}$ for some i, and there are j, l with $R\,k\,l$ and $\varphi \in S^j_{la}$. Hence $\varphi \in S_{la}$, and $S_{la} \supset S_{ka}$ was already established. ⊣

THEOREM 16.3. *A proof-search tree* Tr_{S_1} *for a sequent* S_1 *is closed iff* S_1 *is derivable in LJm*.*

Proof. One direction is evident: A closed proof-search tree is a derivation in LJm*. If Tr_{S_1} is not closed, then it has a non-closed branch. Apply Theorems 16 and 15.0.1. ⊣

References

1. A. Babaev and S. Soloviev. A Coherence Theorem for Canonical Maps in Cartesian Closed Categories. *Journal of Soviet Mathematics*, 20:2263–2279, 1982.

2. H. Barendregt. *The Lambda Calculus.* Number 103 in Studies in Logic and the Foundations of Mathematics. North-Holland, Amsterdam, revised edition, 1991.

3. O. Bittel. Tableau-based theorem proving and synthesis of lambda-terms in the intuitionistic logic. *Lecture Notes in Computer Science*, 633:262–278, 1992.

4. R. Lee C. Chang. *Symbolic Logic and Mechanical Theorem Proving.* Academic Press, New York, 1973.

5. H. Enderton. *A Mathematical Introduction to Logic.* Academic Press, Inc., Orlando, Florida, 1972.

6. M. Fitting. *Intuitionistic Logic, Model Theory and Forcing.* North-Holland, Amsterdam, 1969.

7. R. Gandy. An early proof of normalization by A.M. Turing. In *To H.B. Curry: Essays on Combinatory Logic, Lambda Calculus and Formalism*, pages 453–455. Academic Press, New York, 1980.

8. G. Gentzen. Untersuchungen über das logische Schliessen. *Mathematische Zeitschrift*, 39:176–210, 405–431, 1934.

9. A. Heyting. *Intuitionism, an Introduction.* North-Holland, Amsterdam, 3 edition, 1971.

10. S. Jaskowski. Recherches sur la systeme de la logique intuitioniste. *Actes du Congres International de Philosophie Scientifique*, pages 58–61, 1934.

11. S. C. Kleene. Two Papers on the Predicate Calculus. *Memoirs of the American Mathematical Society*, 10:1–26, 1952.

12. S. C. Kleene. *Mathematical Logic.* Wiley, New York, 1967.

13. S. C. Kleene. *Introduction to Metamathematics.* Wolters-Noordhoff Publishing, Amsterdam, 7 edition, 1971.

14. S. Kripke. Semantical analysis of modal logic. *Zeitschrift für Mathematische Logik und Grundlagen der Mathematik*, 9:67–96, 1963.

15. G. Mints. Axiomatization of a Skolem Function in Intutitonistic Logic. *CSLI Publications*, 2000. To appear.

16. G. E. Mints. *A Simple Proof for the Coherence Theorem for Cartesian Closed Categories*, volume 3 of *Studies in Proof Theory: Monographs*, pages 213–220. Bibliopolis, Napoli, and North-Holland, Amsterdam, 1992.

17. Grigori Mints. Resolution strategies for the intuitionistic logic. In B. Mayoh, E. Tyugu, and J. Penjam, editors, *Proceedings NATO ASI on Constraint Programming, Pärnu, Estonia, 13–24 Aug 1993*, volume 131 of *NATO ASI Series F*, pages 289–311. Springer-Verlag, Berlin, Heidelberg, New York, 1994.

18. H. J. Ohlbach. Semantics-based translation methods for modal logics. *Journal of Logic and Computation*, 1(5):691–746, October 1990.

19. D. Prawitz. *Natural Deduction.* Almquist and Wiksell, Stockholm, 1965.

20. H. Rasiowa and R. Sikorski. *The Mathematics of Metamathematics*, volume 41 of *Polska Akademia Nauk. Monografie matematyczne.* Polish Scientific Publishers, Warsaw, 1963.

21. N. Shankar. Proof search in the intuitionistic sequent calculus. In D. Kapur, editor, *Proceedings 11th Intl. Conf. on Automated Deduction, CADE'92, Saratoga Springs, CA, USA, 15–18 June 1992*, volume 607 of *Lecture Notes in Artificial Intelligence*, pages 522–536. Springer-Verlag, Berlin, 1992.

22. R. Statman. Intuitionistic propositional logic is polynomial-space complete. *Theoretical Computer Science*, 9(1):67–72, July 1979.

23. M. Stone. Representing scope in intuitionistic deductions. *Theoretical Computer Science*, 211, 1999.

24. G. Takeuti. *Proof Theory*, volume 81 of *Studies in Logic and the Foundations of Mathematics.* North-Holland, Amsterdam, 1975.

25. A. Troelstra and H. Schwichtenberg. *Basic Proof Theory*, volume 43 of *Cambridge Tracts in Theoretical Computer Science.* Cambridge University Press, Cambridge, 1996.

26. A. Troelstra and D. van Dalen. *Constructivism in Mathematics, Vol. 1*, volume 121 of *Studies in Logic and the Foundations of Mathematics.* North-Holland, Amsterdam, 1988.

27. A. Troelstra and D. van Dalen. *Constructivism in Mathematics, Vol. 2*, volume 123 of *Studies in Logic and the Foundations of Mathematics*. North-Holland, Amsterdam, 1988.

28. D. van Dalen. *Logic and Structure*. Universitext. Springer-Verlag, Berlin, 3rd, augmented edition, 1994. (1st ed., 1980; 2nd ed., 1983).

Index